8-16-89

MATHEMATICAL VISIONS
The Pursuit of Geometry
in Victorian England

MATHEMATICAL VISIONS

The Pursuit of Geometry
in Victorian England

Joan L. Richards

Department of History
Brown University
Providence, Rhode Island

ACADEMIC PRESS, INC.

Harcourt Brace Jovanovich, Publishers

Boston   San Diego   New York
Berkeley   London   Sydney
Tokyo   Toronto

ACADEMIC PRESS, INC.
1250 Sixth Avenue, San Diego, CA 92101

United Kingdom Edition published by
ACADEMIC PRESS INC. (LONDON) LTD.
24-28 Oval Road, London NW1 7DX

Library of Congress Cataloging-in-Publication Data

Richards, Joan L., 1948-
    Mathematical visions: the pursuit of geometry in Victorian
England / Joan L. Richards.
        p.   cm.
    Bibliography: p.
    Includes index.
    ISBN 0-12-587445-6
    1. Geometry—England—History—19th century.   I. Title.
QA443.5.R53 1988
516'.00942—dc19   87-37349
                                                       CIP

Printed in the United States of America
88 89 90 91   9 8 7 6 5 4 3 2 1

Contents

v

Foreword

Joan Richards believes that mathematics should be recognized as a part of general culture. A number of writers during the past half-century or so have used a historical approach to present mathematics in its historical evolutions and contexts as part of the larger human experience; their number includes such diverse figures as Lancelot Hogben, Morris Kline, and, more recently, Paul Hoffman. But whereas these authors were—to some degree—motivated by the great ideal of giving a general reader some comprehension of the content and significance of technical mathematics, Joan Richards sees her mission to be quite different. Her aim is to display and explain the great change that occurred in British mathematics during the nineteenth century in relation to the great transformations of the conceptual foundations of mathematics, primarily in the area of geometry.

Skillfully interweaving the ideals of British mathematicians, the reward system of the world of mathematics, and the aims of education, Joan Richards introduces us to the ways in which a mathematical research community arises. From her study we learn of the great events in the history of geometry—notably non-Euclidean and projective geometry—and their consequences for the foundations of mathematics. These great advances had important repercussions for teachers, since they altered the established views concerning the nature of mathematical truth or the grounds of knowledge, with major consequences for any consideration of the importance and usefulness of mathematics as a subject in general education.

Joan Richards approaches her task with the insights of a scholar who, in addition to being thoroughly grounded in mathematics, is trained as a professional historian of science. She, therefore, sees the development of mathematics in nineteenth-century Britain from two perspectives in conflict: (1) the shared values of an emerging mathematical community influenced by the changes in the actual subject-matter or content of mathematics and (2) the more traditional ideal of British mathematicians and educators who conceived that mathematics is a subject whose primary purpose is to illuminate the thinking of all intelligent men and women and hence is the principal key to the disciplining and training of the mental functions. Thus, at Cambridge, in the 1830s, the study of mathematics was seen as a means of achieving the humanistic goal of developing the intellect rather than of cultivating the manners of the students. Such a study of mathematics was thus thought to be the counterpart of the study of classics at Oxford. This point of view was based on the notion that "the highest aspiration of life" should be "modelled on a scientist." In these words, William Herschel, in his celebrated "preliminary Discourse," advocated a general scientific education for all, even for those students who would never become professional practitioners in any sense of the word. Yet, whatever their future pursuit might be, whether tending to their estates, pursuing a profession of law or commerce or public life, or becoming clergymen, their scientific education and their understanding of science would unfetter "the mind from prejudices of every kind," leaving it "open and free to every impression of a higher nature which it is capable of receiving." William Herschel expressed in these terms the sentiments of his contemporaries concerning the moral and even religious aspects of learning science and the abstractions of mathematics (especially logic and geometry) on which all sciences ultimately must rest. Herschel concluded the opening chapter of his *Preliminary Discourse on the Study of Natural Philosophy* by discussing how the study of science may "tranquilize and reassure the mind, and render it less accessible to repining, selfish, and turbulent emotions," providing us with a "sense of nobleness and power," enabling us to achieve our potentialities with "strength and innate dignity." The study of science, he concluded, calls forth "those powers and facilities" that form "a link between ourselves and the best and noblest benefactors of our species, with whom we hold communion in thoughts," thereby even participating "in discoveries which have raised them above their fellow-mortals, and brought them nearer to their Creator." Essentially, then, the scientists and mathematicians saw themselves as providing their students with norms of truth and being in fact "the leaders in the human search for truth."

This devotion to truth required the taking of all steps that might possibly advance the understanding, presentation, and use of mathematics. As Joan Richards shows, it was this outlook that led Herschel, Babbage, and others to form an Analytical Society and to transcend the British nationalist tradition of Newtonian mathematics. They successfully campaigned to have their fellow countrymen adopt the more universally useful calculus based on the Leibnizian algorithm, which had come into general use on the Continent.

Joan Richard's explorations of the British views on education and on geometry come to an early climax in her study of the way in which British mathematicians reacted to the new non-Euclidean geometry and its consequent revisions of the concepts of truth. This leads her to the cognate subject of projective geometry and the general revision of views of the nature of mathematical science that occurred in Britain during the third quarter of the nineteenth century. Not only was there a problem between projective and metric interpretations of non-Euclidean geometry, but there were concomitant intellectual struggles concerning the impact of this new mathematics on the nature of the mathematical enterprise and, in particular, on the teaching of elementary geometry in the English schools.

For many readers, the chapter on "Euclid and the English Schoolchild" will be among the most interesting in the book. For here we do not have an ordinary presentation of the methods and content of elementary teaching so much as a discussion of the fundamental ideas of geometry in relation to the training of the mind of British youths. In part, as we see in Joan Richards's portrayal, the discussions hinged on technical questions such as standardized syllabi and examinations. Additionally, however, there was a general concern for such more fundamental topics as the ideals of the training of the mind versus the practical needs of students who might wish to pursue the study of the physical sciences and engineering. In a sense, there arose a conflict between those whose point of view was that of "pure" mathematics, those who saw mathematics (or the mathematical sciences) as purely formal subjects, whose truth depended upon no objective or external referents, and their rivals. The conflict often divided mathematicians into two camps: those who wished to continue the teaching of Euclid in the traditional manner and those who sought to introduce the new geometry and to recognize that geometry itself was no longer a simple unitary concept or a single simple well-defined subject, but one with differing branches, all of which could not be pursued according to identical methods and ideals.

Joan Richards concludes her account with a discussion of the new approaches to the foundations of mathematics at the turn of the century, developments associated in England in the first instance with Bertrand Russell and then with Russell and Whitehead in their collaborative efforts. She stresses the great changes that occurred in the actual practice of mathematicians as a result of the "social and philosophical reorganization of knowledge." She sees the establishment of "a recognized and viable mathematical research community in England" as a force that would change the view of the subject itself. No longer was mathematics to be valued primarily (if not exclusively) for its role in the education of young men and women and its presumed exemplification of the highest form of rationality. The new "ideology" was that of a research community, in which a "mathematician" would be valued not for his ideas concerning education at large or for his ideals of the role of mathematics in education, nor even for his gifts or prowess as a teacher or presenter of the subject at a textbook level. Rather, there was a new career in which the major reward would come from recognition for creative additions to the methods and subject matter of mathematics itself.

This convincing analysis of the changes in the practice of mathematics in Britain in the nineteenth century is presented by Joan Richards in terms of the main currents of thought in philosophy, in education, and in mathematics itself. She shows that, above all, the driving force behind the general arguments was always the set of developments within technical mathematics, chiefly geometry, during the century. This important study whets our appetite for similar explorations located in other countries or cultures. All readers will earnestly hope that such works may be produced so that we may have a more fully rounded picture of the intellectual consequences of mathematical advances.

—I.B. Cohen
1988

Acknowledgements

This book grew out of a doctoral dissertation submitted to the Department of the History of Science at Harvard University. The present version still owes a great deal to the excellent teachers I had there, in particular Erwin Hiebert and I. Bernard Cohen. In addition, Lorraine J. Daston and Shirley A. Roe have been indefatigable supports, offering perceptive criticisms and sage advice through all the stages of its development. I have profited from the responses of many who have heard versions of this work in various guises. Much of the material in chapter three appeared in my article "Projective Geometry and Mathematical Progress in Mid-Victorian Britain," *Studies in the History and Philosophy of Science* 17 (1986):297–325. So many have contributed to the work by listening to my ideas and sharing their insights that to list them would be excessive; they all have my heartfelt thanks.

A variety of people and institutions have supported my work on this book. My dissertation research was supported by a grant from the Harvard Center for European Studies for the summer of 1977, a grant from the Whiting Foundation for the 1978–79 academic year, and a graduate fellowship from the Danforth Foundation. I have drawn resources from the Harvard College Library, the Cornell University Library, the Brown University Library, the British Museum, the Royal Society of London, the Royal Institution in London, the Bodleian Library in Oxford, the Trinity College Library in Cambridge, the University of London Libraries, and the Bertrand Russell Archives at McMaster University.

Introduction

In the second half of the nineteenth century, a number of apparently specific scientific innovations were suggested which fed into massive changes in Western intellectual life. Perhaps the most visible example of such a new development is Darwin's theory of evolution. With his speculations about the origins of biological species, the reclusive Englishman raised questions about the nature of history, society, human nature and human knowledge which are still reverberating through Western culture. More esoteric perhaps, but in many respects equally powerful, were developments in mathematics. This is particularly clear in the study of geometry, where new ideas cast serious doubt on traditional views of the nature of the conclusions that could be reached through mathematical study. Echoes of these doubts and changes were heard across the board, in philosophy, education, art, physics and even law.

New geometrical ideas had this powerful and far-reaching impact because of the central importance of the Euclidean geometry they were expanding or displacing. After the triumph of the Newtonian system of the world, geometry was seen as the most dramatically successful science. Most nineteenth-century philosophers perceived Euclidean geometrical axioms as descriptions of the fundamental properties of spatial reality. These axioms, and the theorems generated deductively from them, had a twofold character. On the one hand they were indubitably true. The axioms were so obvious that

1

once they were properly understood, no one could seriously doubt them; once they were accepted, the demonstrations of more complicated theorems required merely that one be able to think clearly and accurately. From a subjective point of view, then, geometry was unassailable.

At the same time, geometry was objectively impeccable. Whereas other theoretical systems might be defended as well thought out, they were relatively weak empirically; a cushion of approximation separated the theoretical from the experiential situation. Euclid's geometry, on the other hand, was exactly true in the objective world; the properties of physical circles, triangles and of space itself were precisely what the theory predicted. Triumphantly accurate in both the subjective and the objective realms, geometry was the *summum bonum* of human knowledge. It was truly the queen of the sciences.

In the post-Newtonian world, the indubitability and exactness of geometrical truth were seen to be unique among man's scientific insights. With geometry, humans seemed to have bridged the Cartesian gap between mind and body, to have transcended the confines of subjective being and attained a true understanding of the external world. However, the existence of this knowledge raised the critical question of how it had been acquired. The extraordinary success of geometry forced all who tried to formulate a theory of human learning to confront the issue of how people could come to have this kind of exact knowledge of absolute truth.

People responded to this question in a variety of different ways. Kant's interpretation of space as an essential category of human knowledge was perhaps the best known and most influential solution for the nineteenth century. However, although the German's treatment was so powerful it often dominates considerations of geometrical epistemology, Kant was far from alone in recognizing that understanding the nature of geometry posed a fundamental challenge for attempts to understand the limits and extent of what could be known. He was one of a long line of epistemological inquirers— including Locke, Berkeley, Hume, Reid, Stewart and their intellectual heirs—all of whom grappled with the nature of geometrical truth.

The epistemological theories these men developed are often classified as either nativist or empirical according to their treatment of the way humans gain knowledge. Interpretations of geometry were central to this division. Philosophical nativists argued that true knowledge of the natural world was in some essential way innate; in the last analysis things were known intuitively rather than being built up from experience. The deductive power and exactness of the geometrical knowledge of space was a touchstone for their arguments that knowledge of the essential structures of the natural world were inherent in our minds, ready to unfold if rightly tapped.

This same geometry was a stumbling block for empiricists who claimed that natural knowledge was not innate but instead was drawn from experience in the external world. This position raised the question of how absolutely true geometrical knowledge could be generated from the same inductive processes which led to relatively inconclusive and inexact results in other areas. Empiricists were hard-pressed to explain how we could conclusively prove the truth of statements about space when proving even the simplest physical statements was not only laborious but inexact as well. The issue was particularly pressing when one considered theories like that of parallels, where geometers seemed to be making true statements about infinite phenomena. An empirical theory of knowledge would suggest that valid knowledge is bounded by experience, and yet we have no experience of lines produced indefinitely in the far reaches of space. Euclid's definition of parallel lines asserted knowledge of how these lines would behave, however. Universally acknowledged to be true, such geometrical statements posed huge problems for nineteenth-century empirical epistemologists.

Thus, geometry focused much nineteenth-century discussion of the nature of human knowledge. Philosophers defending either the nativist or the empirical positions, as well as those whose ideas lay somewhere in between, returned to the subject constantly because it forced them to consider how humans could come to exact knowledge of absolute truth. Since the epistemology of this era inevitably involved incorporating the peculiar knowledge of geometry, classifications of knowledge almost without exception placed the subject in a central position, and discussions of man's intellectual aspirations and limits consistently referred to it as a pivotal example.

A number of nineteenth-century developments in geometry shook the common ground on which the discussions between nativists and empiricists were staged. The so-called non-Euclidean geometries based on different theories of parallels were perhaps the most startling development. Additionally, the notion of spaces of more than three dimensions opened floodgates of speculation about whether human experience was necessarily bounded by Euclidean terms. Less radical, perhaps, but equally novel, were unconventional approaches to Euclidean space; projective geometry generated a whole new set of theorems and relationships which were previously unsuspected. All of these were new developments in an area which always before had been circumscribed by Euclid's *Elements*. Their cogency undermined the conviction that Euclidean geometry was the definitive presentation of a completely known form of knowledge. In different ways, all of these geometrical developments created challenges for the carefully constructed nineteenth-century view of geometry as a subject wherein subjective and objective worlds were seamlessly joined in complete accord.

Therefore, these often interrelated geometrical developments engendered a series of discussions about the subject of mathematical study, about the relation of mathematics to the reality studied in the natural sciences and about the nature of human knowledge in general. In the long run these discussions culminated in a radical change in the perception of the nature of geometry. Whereas in the nineteenth century, geometrical results were perceived descriptively, as binding truths about real space, in the twentieth they are more commonly seen formally, as deductions drawn from an abstract axiom system only more or less applicable to the real world.

The distinction between these two views can be illustrated by considering the specific example of a circle. In geometry, a circle might be defined as the locus of points which lie equidistant from a given point. From this definition, combined with a series of axioms and postulates, one could develop a mathematical system which would elaborate more of the properties of circles. In this basic approach, nineteenth-century geometry is like twentieth-century geometry which in turn is like the geometry of the Greeks.

The point of difference concerns the relationship between the circles which are constructed through this kind of mathematical development and the circles which we encounter as wheels on carts, compass drawings on paper or visions in the mind's eye. In one interpretation, the "descriptive" interpretation, the circles which are the focus for mathematical discourse are intimately tied to the circles encountered elsewhere—whether as physical, visual or intellectual objects. In this view, a circle is a circle no matter how Euclid or any other geometer might wish to speak of it; in the final analysis, their mathematics is right or wrong depending on how accurately their circles accord with the circles which are encountered in other contexts.

In contrast to this descriptive view is the "formal" view. The formal view focuses attention on the system itself as the ultimate determinant of mathematical validity. In this view, the only thing which establishes the validity of a mathematical system is that all of the objects—whatever they are labeled—be defined and treated in such a way that their properties don't conflict with those of other defined objects. As long as the system which is mathematically unfolded does not contradict itself internally, it is considered to be mathematically true. In contrast to the descriptive view, from the formal perspective the relationship between mathematically constructed sytems and non-mathematical experience, be it physical, visual or conceptual, is irrelevant to mathematical truth. Mathematical circles bear no essential relationship to circles encountered outside of the subject; it is in an important sense by accident that a mathematical object, like a circle, shares its name with circles encountered in other spheres of life. Whatever the circumstances which might lead one to identify circles encountered outside of the classroom with those developed within it, this correspondence is not essential to the truth of the mathematical development.

As here defined, the terms descriptive and formal are very broad; a wide variety of very disparate views of mathematics can be grouped under each of them. They are useful, however, for organizing the rich and complex historical situation in which geometry developed in the nineteenth century. That development was a long and often slow process. As the century progressed, a number of different traditions encouraged different emphases within the broad area now described by the term non-Euclidean geometry. Furthermore, at the same time, other geometrical theories, like projective geometry, raised important and closely related issues. Not only did new mathematical ideas generate new perspectives on the nature of geometry, but changing traditions in philosophy and psychology suggested new approaches. Geometry in the nineteenth century did not develop as an unbroken line of new ideas, each generated inexorably from the one preceding it, but was rather like the growth of a river, or perhaps a lake, fed by numerous different streams of thought.

This book is a study of changing views of geometry in the second half of the nineteenth century, when the silence which initially greeted Lobachevskii's and Bólyai's works was replaced by a cacophony of diverse voices. In it, geometry is defined broadly, in the way that its nineteenth-century developers approached it. Therefore the treatment of geometry is not confined to technical works and new mathematical developments but encompasses the philosophical, psychological and educational discussions of the subject which were carried on concurrently. The aim is to trace the changing interpretations of geometrical truth in relation to a broader cultural context rather than to criticize, evaluate or interpret the ideas being propounded.

This historical, as opposed to mathematical or philosophical, approach affords the opportunity to clear up some anomalies which have plagued other investigations in the form of apparent blind spots in nineteenth-century vision: the long lapses between ideas introduced and their development, the dead ends enthusiastically preferred to what in hindsight were more fecund interpretations. These situations are anomalous only in the context of the internalist and progressive framework that has long been used to view the development of mathematics in the nineteenth century. They can be exorcised by creating a broader framework that includes heretofore disregarded forces which acted on those who were engaged in these developments. Taking this broader historical view involves maintaining that extra-mathematical factors, such as national culture and social situation, played important roles in the discussions about about geometry at the end of the nineteenth century.

In this work I am focusing primarily on the reception and interpretation of the new geometrical theories in England. The choice of England as the focus of the study was dictated partly by the merely pragmatic consideration that having vastly broadened the definition of mathematics, the historical situation needed to be narrowed if it was to be approached at all. England defined a relatively small and homogeneous mathematical community with which to work. Equally important was a peculiar aspect of late nineteenth-century English culture. In England, the essential connection of mathematics to all other aspects of human development was a constant theme, not only reiterated frequently in philosophical musings on mathematics but institutionalized firmly in the educational system and accepted implicitly in notions like that of the mathematically literate gentleman. Almost half of England's educated elite, those who took their degrees at Cambridge, had to be proficient in mathematics. Though passing the Mathematical Tripos ceased to be a requirement for all honors degrees in the 1850s, throughout the century the rigorous mathematical training necessary to place well on this examination continued to attract large numbers of students every year—students who then went forth to pursue careers in virtually every facet of English life.

The size and intellectual vigor of the group who, in the nineteenth century, struggled through the Mathematical Tripos at Cambridge contrasts starkly with a noticeable dearth of English mathematical creativity at the same time. It is almost a truism in the history of mathematics that in the nineteenth century most interesting developments in pure mathematics took place *outside* of England: in France, Germany or Italy. Considering the number of Englishmen who seriously studied the subject, this seems strange.

The apparent discrepancy between the educational output and the mathematical creativity of nineteenth-century England can be explained by recognizing that on an important level, mathematics was not seen as a specialized, research subject but rather as a universal, educative one. Its pride of place in the Cambridge curriculum was defended in terms of its central exemplary position among fields of human knowledge. Its generalizability constituted its attraction; those who could think mathematically could think effectively. Pure mathematics, even at its highest levels, was more often defended as intellectual training than as an autonomously valuable specialized study.

This situation may not have encouraged mathematical research, but it provides a fertile field for considering nineteenth-century geometrical issues. The English mathematical education was a concrete institutional manifestation of the epistemological assumptions which surrounded the study of geometry. For the historian, it provides a convenient indicator of the status of these views, as well as a forum in which they were carefully discussed. In the educational context, the abstract musings of philosophers were brought to earth, and their relationship to the historical realities of English life were explicitly negotiated.

The peculiar position of mathematics in English higher education is not merely useful as a window for the questing historian; it played an important role in the historical development of English mathematics itself. The educational equation of mathematical thinking with sound reasoning meant that in nineteenth-century England, the study of mathematics was deeply and explicitly enmeshed in a whole tangle of broader intellectual and cultural issues. The impact of geometrical developments that carried foundational implications ranged from the ideological concerns of educators to the philosophical concerns of theologians to the social concerns of parents wanting the best education for their children. The swirl of implications attendant on new geometrical theories is obvious and explicit during this period. This makes nineteenth-century England a particularly good field in which to explore the ways mathematical development interacts with its cultural context.

The first chapter of the book traces the basic parameters of the mid-century English view of geometry, considering its philosophical, institutional and mathematical manifestations and implications. The interpretation which embedded the study of geometry deep in English views of knowledge and the human condition was essentially descriptive. This approach formed a link between geometrical knowledge and the knowledge developed in scientific fields, where theories also described external objects. Some form of the descriptive approach to geometry, with its close ties to science, formed the background for the dynamic of geometrical and intellectual change which was played out through the rest of the century.

The second chapter considers the mid-century development of non-Euclidean geometry. In the 1860s there was a flowering of the subject on the continent. Lobachevskii's and Bólyai's works were rediscovered and eagerly perused. These were supplemented by new works by two Germans: Bernhard Riemann and Hermann von Helmholtz. The subject finally came into its own. At the same time, it came to England. Continental non-Euclidean ideas were first recognized and discussed in England in the 1860s and 1870s. They introduced specific tensions and strains in the English view of the nature of mathematics, of geometry in particular, which were acutely uncomfortable for many of the country's intellectuals. The suggestion that more than one geometry might describe our spatial experience directly threatened the descriptive interpretation of mathematical truth which was so central to English perceptions of geometry and all of those subjects linked to it.

The third chapter shifts to a different mathematical development, projective geometry. This subject originated in France in the first decades of century and began to draw English attention in the 1840s. By the 1870s it was widely hailed and pursued as "the modern geometry." Part of its appeal in the final quarter of the century lay in the fact that this study provided a radically different interpretation for non-Euclidean geometries, which significantly altered their meaning in England. Projective geometry provided an exciting compromise which allowed new work to be carried out within a descriptive interpretation of mathematical foundations. Thus, the enthusiastic pursuit of projective geometry in late nineteenth-century England was not merely inchoate enthusiasm for a new field, but a response to a much broader set of intellectual issues in which geometry was involved. The English excitement over projective geometry reveals the major features of their ever-developing view of geometry in the final decades of the century.

Although, when viewed descriptively, projective geometry provided a more palatable interpretation for the new, non-Euclidean theories of parallels, it shared an important characteristic of that work; it was not Euclidean. This was a significant issue. In England the boundaries of elementary geometry were very clearly defined as Euclid's *Elements*. It was geometry as set forth in this book that was universally recognized as the epitome of sound reasoning; as such, the *Elements* itself was the standard textbook in the secondary schools. This emphasis on Euclid was reexamined and renegotiated at the end of the century. Though the mathematical issues were relatively elementary, the late-century discussions of Euclid as a textbook also raised serious questions about the fundamental nature of geometry. Thus, the changing fortunes of the *Elements* of Euclid in late nineteenth-century England is the subject of the fourth chapter.

The fifth chapter focuses on the 1890s. It returns to considerations of the meaning assigned to the most modern developments in geometry, both projective and non-Euclidean. By the final decade of the century, the mid-century descriptive understanding of geometry and its place in human knowledge had become rather worn. The basic features of this interpretation which had endured for so long were still in place, however, and continued to invite exploration. The development of Bertrand Russell's earliest interpretation of geometry serves to organize this chapter's consideration of the final configuration of the nineteenth-century tradition.

The epilogue points out the emergent features of the twentieth-century view of geometry, which, in the first decade of the new century, replaced the descriptive approach with startling rapidity. The new interpretation of geometry was so radically different from that which had gone before that it obliterated interest in much of what had concerned nineteenth-century geometers. When they embraced the new point of view, mathematicians changed their subject so radically that to pursue its development past this point would require a second book.[1] Therefore, the establishment of a formal view of mathematics provides a quick and natural end to this study.

This book originated as a study of the nineteenth-century reception of non-Euclidean geometry in England. As the above chapter outline suggests, however, the attempt to do justice to the complexity of that reception has led in so many directions that the original focus seems artificially narrow. What began as a consideration of a

---

[1] This assertion raises the philosophical question of whether there have been revolutionary changes in mathematical history akin to those Thomas Kuhn claimed for science in his classic, *The Structure of Scientific Revolutions,* (Chicago: The University of Chicago Press, 1962). Kuhn exempts mathematics from his model, as have many observers of mathematics since (viz. *Historia Mathematica,* 2 (1970):437–72). Recently however (see, for example, Philip Kitcher, *The Nature of Mathematical Knowledge* (New York: Oxford University Press, 1983), the exempt status of mathematics has been seriously questioned. (For an excellent compilation of articles on all sides of this issue, see Thomas Tymoczko, *New Directions in the Philosophy of Mathematics,* (Boston: Birkhauser, 1985).) I do not wish to address the detailed philosophical issues involved in this discussion here but hope that this study may provide historical material useful for those who do.

highly refined mathematical development has become a broader study of mathematics in culture, focusing on the pursuit of geometry in Victorian England.

The reason for this is simple. Non-Euclidean geometry derived its interest and impact from its foundational implications. It raised fundamental questions for long-standing interpretations of the nature of geometrical study. These interpretations can be found most easily in erudite philosophical texts or in the writings of mathematical practitioners. They are usually regarded, therefore, as essentially recondite and esoteric issues, which are addressed only on these specialized levels.

Ultimately, however, foundational questions can be reduced to the question of what mathematics is. This question is at once simple and immense. It is not relevant only to the systems of philosophers or even of mathematicians. It is addressed over and over again as societies negotiate the terms of mathematical education, development, support and prestige. Responses to it are better understood as visions than as answers; they appear as ever-changing patterns rather than fixed solutions. To borrow a phrase from an English mathematician: "like a rainbow, if we try to grasp it, it eludes our very touch; but like a rainbow it arises out of real conditions of known and tangible [circumstances.]"[2]

This study portrays the mathematical visions of the Victorians. Such a portrayal gives coherence to their mathematical interests. At the same time, it provides a view of the society which supported those interests. It explores for a particular time and place—at once hauntingly similar and essentially different from ours—the ways in which mathematical ideas have been woven into the very fabric of Western culture.

---

[2] William Spottiswoode, "Presidential Address," *Report of the Forty-eighth Meeting of the BAAS held at Dublin in August 1878,* (London: John Murray, 1879), p. 21. The full quotation is on p. 58.

John Herschel. *(Courtesy of the National Portrait Gallery, London, England.)*

William Whewell. *(Courtesy of the National Portrait Gallery, London, England)*

John Stuart Mill. *(Courtesy of the National Portrait Gallery, London, England.)*

Augustus De Morgan

CHAPTER 1

The Mid-Century View of Geometry

The year 1818 may be taken as a convenient starting date for the nineteenth-century chapter of British mathematical development. This is the year when Charles Babbage, John Herschel and George Peacock, all part of a group of mathematical aficionados calling themselves the Analytical Society, published their English translation of a French calculus text in hopes of introducing French mathematics into Britain. Within a year, the format of the Cambridge mathematical examination, also called the Senate House Examination or the Tripos, was changed to allow the use of non-Newtonian symbology in solving problems. The people who advocated moving away from Newton's approach to the calculus saw the change as part of a broader movement to encourage the pursuit and development of science and mathematics in England.

The group of men who are first identifiable as members of the Analytical Society went on to become the core of English science for the first half of the nineteenth century. Charles Biddel Airy, George Peacock, Charles Babbage, John Herschel, William Whewell all seem to be ever-present in English scientific circles. They exemplified scientific activity in teaching and research, they advocated the study of science in a variety of publications, and they played pivotal roles in scientific societies large and small. Their commitment to the advancement of science in England was unstinting, and

13

by the time they died—most in the 1860s—they had left their mark on virtually all scientific subjects.

The mark they left was a humanizing one; the science they supported was an integral part of human existence. In his book *A Preliminary Discourse on the Study of Natural Philosophy*, first published in 1830, John Herschel eloquently stated this position. "Man is constituted a speculative being," he wrote;

> he contemplates the world, and the objects around him, not with a passive, indifferent gaze, as a set of phenomena in which he has no further interest than as they affect his immediate situation, and can be rendered subservient to his comfort, but as a system disposed with order and design. He approves and feels the highest admiration for the harmony of its parts, the skill and efficiency of its contrivances. . . . Thus he is led to the conception of a Power and an Intelligence superior to his own, and adequate to the production and maintenance of all that he sees in nature.[1]

Herschel advocated scientific study as an essential part of a comprehensive spiritual and intellectual quest. Scientific knowledge is good for the soul, he claimed. It should be widely and freely disseminated throughout society for it calms, uplifts, tranquilizes and reassures the mind.

> And this it does, not by debasing our nature into weak compliances and abject submission to circumstances, but by filling us, as from an inward spring, with a sense of nobleness and power . . . by showing us our strength and innate dignity, and by calling upon us for the exercise of those powers and faculties by which we are susceptible of the comprehension of so much greatness.[2]

Herschel's ennobling interpretation of the function of scientific knowledge in culture was shared by many of England's scientific advocates. Cannon has emphasized their position as a balance to the Oxford Movement by labeling this group "The Cambridge Net-

---

[1] John Frederick William Herschel, *A Preliminary Discourse on the Study of Natural Philosophy,* (London, 1830: rpt. New York: Johnson Reprint Corporation, 1966), p. 4.
[2] Ibid., p. 16.

work."[3] Morrell and Thackray have focused on the civilizing aspects of their view by referring to them as "Gentlemen of Science."[4] Schweber has found their integration of science with other knowledge to be central and called them "Scientists as Intellectuals."[5] Whichever way one characterizes this group, the cultural interpretation of scientific study, so eloquently stated by Herschel, was their answer to those who attacked science as a threat to religion, who feared it as it as dangerous to cultural stability or who denigrated it as merely a way of catering to pampered intellectual or industrial appetites. In the view Herschel articulated, scientific knowledge was not to be associated automatically with pragmatic problems or the dirt, grime and social unrest of factories. On the contrary, it was an integral part of a humanizing cultural legacy.

The institutional framework for this kind of humanistic science in England is often hard to pin down. Scientific work was carried out in a bewildering array of institutional contexts with styles ranging from the small, specialized intensity of the Geological Society in London, to the genteel exclusiveness of the Royal Society, to the circus atmosphere of the British Association for the Advancement of Science (BAAS), sometimes referred to as the British Ass. Some of the men initially identifiable by their involvement with the Analytical Society were active in all of these spheres.

Mathematics held a somewhat special position in this picture. The London Mathematical Society was not founded until 1868, by which time the men initially identifiable in relation to the Analytical Society were too old to play a sustained role. In the Royal Society and the BAAS, pure mathematics was not particularly encouraged; in the latter organization it shared section A with physics throughout the century. In short, as an autonomous study, pure mathematics was rather neglected.

---

[3] Susan Cannon, *Science in Culture: The Early Victorian Period,* (New York: Science History Publications, 1978).

[4] Jack Morrell and Arnold Thackray, *Gentlemen of Science: Early Years of the British Association for the Advancement of Science,* (Oxford: Clarendon Press, 1981).

[5] S. S. Schweber, "Scientists as Intellectuals: The Early Victorians," in James Paradis and Thomas Postlewait eds., *Victorian Science and Victorian Values,* (New York: New York Academy of Sciences, 1981), pp. 39–68.

Nonetheless it was a subject of central importance for middle-class England's scientific outlook. The place where Victorian mathematics was most consistently encouraged was in education. A large proportion of England's scientific enthusiasts of the late nineteenth century were products of the mathematical education at Cambridge. The members of the Analytical Society are initially identifiable as students there: it was the jumping off place for their lifelong attempts to revitalize and mold the study of science in England. Other areas of science may have caught their attentions as they matured, but it was on the rock of Cambridge mathematics that they constructed their views of what these sciences entailed. Almost thirty years after their initial appearance on the English intellectual scene many of the same men were again directly involved in reevaluating the Cambridge curriculum and revising the Tripos. The particular event around which their considerations of mathematical knowledge coalesced was the Tripos Reform of 1848.

The Analytical Society's earliest efforts were felt on the Tripos in 1819. At that time, when the young George Peacock was moderator, Newton's fluxional notation was abandoned on the Senate House Examination. In the ensuing years this innovation took hold, and analytical notations were universally adopted. Abandoning the geometrical, Newtonian notation of the previous century was advocated as a significant step towards bringing English scientists into the same universe of discourse as their continental counterparts. It marked the rise of a new analytical emphasis at Cambridge and an attempt to include all of the most up-to-date developments in mathematics, both pure and applied, on the Tripos.[6]

---

[6] This development is mentioned as a central feature of nineteenth-century British mathematics by virtually everyone who has considered it. The classic treatment is in W. W. Rouse Ball, *A History of the Study of Mathematics at Cambridge,* (Cambridge: University Press, 1899). More recent studies of this reform and its implications include J. M. Dubbey, "The introduction of the differential notation to Great Britain," *Annals of Science,* 19(1963), 37–48; Maurice Crosland and Crosbie Smith, "The transmission of physics from France to Britain: 1800–1840," *Historical Studies in the Physical Sciences,* 9 (1978):1–61; Philip C. Enros, "The Analytical Society: Mathematics at Cambridge University in the Early Nineteenth Century," Toronto, 1979, unpublished PhD dissertation; Philip C. Enros, "The Analytical Society (1812–1813): Precursor of the Renewal of Cambridge Mathematics," *Historia Mathematica,* 10 (1983):24–47.

However, the development of the Tripos did not stop in the 1820s. In the wake of the liberating reforms the purview of the examination expanded unmanageably. Since any aspect of mathematics might be included on the Tripos, it became virtually impossible to prepare for it. In the middle of the century, as part of an attempt to define and delimit the subjects appropriate to the examination, the place of mathematics in the Cambridge curriculum was searchingly reexamined and reevaluated. Many of the now aging members of the Analytical Society were again involved in these deliberations. An exceptionally clear picture of mid-century English views of the nature of mathematics and its relationship to intellectual culture at large emerges from the discussions which culminated in the 1848 reform. These views form the basic backdrop for virtually all considerations of mathematical, more specifically geometrical, innovation throughout the rest of the century.

In this chapter, the basic features of this mathematical scenery will be presented, starting from the widest philosophical considerations and moving to more specific mathematical issues. In the first section, the basic terms of the philosophical view of mathematics will be presented. This view, which was largely hammered out in the context of educational concerns, defined the basic supporting foundation for mathematical work throughout the century. Any attempts to alter its basic parameters had profound institutional and intellectual implications which were generally recognized. As will become clear in subsequent chapters, whenever changes were suggested, they were fundamentally important and engendered passionate response.

In the second section, the magnification will increase while the focus is narrowed to show the ways in which these most general perceptions of mathematics and its relation to intellectual culture as a whole related to hierarchical distinctions within mathematics itself. In 1818, the members of the Analytical Society had advocated the study of analysis in an attempt to move away from the geometrical emphasis of the Newtonian legacy. This position was reversed in 1848 when the same men were important actors in a move to reemphasize geometry as the core of the Cambridge liberal education. The considerations which led to this change in position point to a number of issues which were later renegotiated yet again with respect to new developments on the forefront of geometry.

Mathematics itself was not wholly quiescent in the period directly following the Tripos reform however. Even as strictly Euclidean geometry was being canonized on the examination, new developments were challenging its adequacy. The third section will explore some of the practical problems the accepted views of geometry were creating at the forefront of research. To continue the microscopic analogy, the lens here will be stronger, the focus finer yet, to show the ways that the broadest attitudes towards the position of mathematics in intellectual culture impinged on detailed questions generated within the subject itself.

MATHEMATICS, REASONING AND TRUTH IN MID-CENTURY ENGLAND

Mathematics at Cambridge was pursued within the context of a liberal, as opposed to a specialized, education. The English liberal education was ostensibly designed to broadly educate students for civilized life. In the eighteenth century, this goal had been pursued through the development of a certain standard of behavior in the liberally educated man. In the more dynamic and competitive atmosphere of the nineteenth century, the emphasis shifted from the development of manners to the development of intellectual skills which the educated man could apply to the varied exigencies of adult life. The change reflects important differences in the English experience in the two centuries.

The nineteenth-century intellectual emphasis in education was accompanied by an increased focus on written examinations as a way to determine a student's rank. As the century progressed, examinations became increasingly central to the evaluation of students at the English Universities, and their education was increasingly directed towards success on crucial tests. This was true both at Cambridge and at Oxford.[7]

---

[7] For a fuller study of changing perceptions of the liberal education see Sheldon Rothblatt, *Tradition and Change in English Liberal Education* (London: Faber and Faber, 1976).

The unique aspect of the education at Cambridge, which set it off from Oxford, was that the major emphasis of its curriculum was upon mathematics. Until the 1850s, no matter what subject formed a student's primary interest, he had to study mathematics to obtain an honors degree. At Oxford, the primary focus of the education was on the classics. But at Cambridge, even if Greek and Latin were his major interest, a student could not take the classics examination without first passing the Mathematical Tripos.[8]

The mathematical emphasis of the Cambridge curriculum was justified as a part of a liberally educating course of study. In the middle of the century a professional need for mathematical training, even for those intending engineering careers, was not generally acknowledged. Although this situation was beginning to change in the middle of the century, the major context in which serious mathematical study was institutionally encouraged continued to be the liberal education at Cambridge. Even the growing strength of mathematics at University College in London did not significantly change it. Augustus De Morgan justified his teaching there in ways which revealed a similar spirit to that at his *alma mater*. Furthermore, many of his best students, like J. J. Sylvester, went on to study at Cambridge.

An important effect of this educational emphasis is tellingly revealed in the "Report" of a Parliamentary Commission appointed in 1852 to investigate the Cambridge education. While extolling the virtues of mathematical studies for the progress of civilization, the Commissioners noted:

> There were danger that Mathematics might perish from the face of the earth if not made an especial part of a liberal education, and that at all events their higher forms would never arise unless their attainment were cherished and fostered by high rewards, . . . and unless some powerful encouragement were afforded for a continuance in pursuits of all others the least lucrative in themselves.[9]

---

[8] For a lively, first-hand account of the effects this had on students primarily interested in the classics see Charles Bristed, *Five Years in an English University*, 2 vols. (New York: Putnam, 1852).

[9] Great Britain, Parliament, *Parliamentary Papers*, 1852–53, vol. 44 (*Reports*, vol. 5), Cmnd 1559, "Report of her Majesty's Commissioners appointed to inquire into the State, Discipline, Studies, and Revenues of the University and Colleges of Cambridge: together with the Evidence and an Appendix." From "Report," p. 105.

Clearly pure mathematics was in a precarious position. In fact, the Commissioner's statement leads immediately to the question of why a study with such tenuous attractions should be pursued at all. This question was actively debated in the 1830s and 1840s. During this period, William Whewell, formerly allied with the Analytical Society, led the defense of Cambridge mathematics against those who alleged it was too narrow and esoteric to be educationally useful. The key to Whewell's defense was a descriptive view of the nature of mathematics with which he countered a more formal view espoused most notably by some Scottish philosophers.

Although as a philosopher Whewell was a nativist, and therefore often at odds with his more empirically oriented compatriots, on critical issues concerning mathematics his views were widely shared. The descriptive view of foundations which he clearly defined and fiercely defended in the 1830s and 1840s formed the bedrock of Cambridge mathematics throughout the century. On this issue his empirically minded friend, John Herschel, agreed with him. Perhaps more striking, even Whewell's intellectual archenemy, John Stuart Mill, who had no particular ties to the education Whewell was defending at Cambridge, agreed with the descriptive emphasis in mathematics. Obviously, even three men as influential as Whewell, Herschel and Mill did not completely prescribe a single view of mathematics which was embraced by all of their compatriots. However, their agreement suggests the central importance of the point which drew them all together. Whatever else might have divided them, mid-century English philosophers approached mathematics as a descriptive study.

Mathematics and Reasoning: William Whewell and William Hamilton

The descriptive interpretation of mathematics was clearly set out in Whewell's early writings about mathematics in education. In 1835, he examined and defended the mathematical emphasis of the Cambridge curriculum in a pamphlet entitled "Thoughts on the Study of Mathematics as a Part of Liberal Education." Here Whewell presented the educational issue as follows:

> This complete mental culture [which was to be imparted through liberal education] must, no doubt, consist of many elements; but it is

certain that an indispensable portion of it is such a discipline of the reasoning power as will enable persons to proceed with certainty and facility from fundamental principles to their consequences. And this part of the cultivation of the mind is what I shall at present especially consider. Let us suppose it established, therefore, that it is a proper object of education to develop and cultivate the reasoning faculty. The question then arises, by what means this can be done;—what is the best instrument for educating men in reasoning.[10]

The answer Whewell offered was mathematics. It was through mathematical study, he claimed, that students could learn to reason effectively in all areas of their adult lives.

Implicit in Whewell's advocacy of focused mathematical study as the backbone of a generally educating curriculum was a basic assumption about the nature of the human mind which he shared with most of his contemporaries. The entire search for a subject matter whose study could teach youth reasoning in general rested on the assumption that essentially the mind was a conglomeration of intellectual skills, usually called *faculties,* which could be strengthened by intellectual exercise. Thus, Whewell argued for the inclusion of subjects in the curriculum in terms of the widely applicable faculties which such a study would strengthen. It was because of its power to train the faculty of reason rather than its value as a useful study *per se* that Whewell argued for the study of mathematics.

In emphasizing that the aim of University study was the development of widely applicable mental faculties, Whewell was arguing within an educational philosophy which was widely assumed in England. In deciding that mathematics was the optimum subject for developing the skills necessary for "complete mental culture," he was taking a further stand. He was claiming that the study of mathematics—as opposed, for example, to the study of logic—would most effectively strengthen basic reasoning skills. In Whewell's view, a mathematically educated person would make a good lawyer, parliamentarian, doctor or clergyman because his studies would have taught him how to think effectively.

---

[10] William Whewell, "Thoughts on the Study of Mathematics as a Part of a Liberal Education," Appended to *On the Principles of English University Education,* (London: John W. Parker, 1838), p. 139.

The Scottish philosopher William Hamilton sharply attacked this position on the grounds that mathematics was too unlike other subjects to have any broadly educating effects. For Hamilton, mathematics was an essentially deductive study which had but few affinities with the substantial thinking of other intellectual endeavors. Whereas scientific theories were descriptive, mathematical ones were formal. Theorems generated deductively from mathematical axioms were not substantially true, Hamilton claimed, but merely the logical consequences of a hypothetical system. In practical life, on the other hand, one did not reason deductively from such principles but rather directly on a variety of subjects. In Hamilton's view, then, mathematics was different from other studies because it was formally based. As such it was an empty and isolated study; hence it had virtually no value as a part of the liberal education.[11]

Whewell responded to this kind of criticism in "Remarks on Mathematical Reasoning" published in 1837. He here challenged the view, which he attributed to Hamilton's compatriot, Dugald Stewart, "that mathematical truth is hypothetical, and must be understood as asserting only, that *if* the definitions are assumed, the conclusion follows."[12] Instead he argued for a highly descriptive, conceptual view of mathematics. In his treatment the crux of the issue separating the formal and descriptive views lay in considerations about the nature of mathematical definitions.

As Whewell saw it, the Scottish position was that mathematical definitions are, *au fond*, arbitrary. They serve as the basis for intellectual systems which are deductively constructed but do not necessarily have a real or conceptual subject matter at all. Whewell, on

---

[11] [William Hamilton], "Review of Thoughts on the Study of Mathematics as a part of Liberal Education," *Edinburgh Review*, 62 (1836):218–52; William Hamilton and William Whewell, "Note to the Article on the Study of Mathematics," *Edinburgh Review*, 63 (1836):142–44. The Scottish school of science and its relations to England have long fascinated scholars. For the situation in Scotland see George Elder Davie, *The Democratic Intellect: Scotland and her Universities in the Nineteenth Century*, (Edinburgh: University Press, 1961). For English-Scottish relations in science see Richard Olson, *Scottish Philosophy and British Physics, 1750–1880: A Study in the Foundations of Victorian Scientific Style*, (Princeton: Princeton University Press, 1975); Richard Olson, "Scottish Philosophy and Mathematics, 1750–1880," *Journal of the History of Ideas*, 32 (1971):29–44.

[12] William Whewell, *The Mechanical Euclid, to which are added Remarks on Mathematical Reasoning and on the Logic of Induction* (Cambridge: J. and J. J. Deighton, 1837), p. 147.

the other hand, argued that definitions are inextricably tied to the object being described. It is impossible, Whewell asserted, to create purely arbitrary definitions. Using as his example the definition "A line is said to be *straight* . . . when two such lines cannot coincide in *one* point without coinciding altogether," Whewell pointed out that:

> It would inevitably be remarked that no such lines exist; . . . or, more generally, that the definition does not correspond to any conception which we can call up in our minds and therefore can be of no use in our reasonings. And thus it would appear, that a definition, to be admissible, must necessarily refer to and agree with some conception which we can distinctly frame in our thoughts.[13]

For Whewell then, mathematical axioms and theorems were not merely formal statements; they were descriptive. Their truth rested in their fidelity to the complicated primitive concepts they described: the concepts of number, space, etc. The essence of mathematical knowledge was not attained merely by understanding the deductive connections among terms but required a total understanding of the things, like straight lines, to which those terms referred.

The bulk of Whewell's arguments from "Remarks on Mathematical Reasoning" was incorporated in his major work *The Philosophy of the Inductive Sciences*.[14] Its position in this work extended the impact of his arguments about the nature of mathematics far beyond the confines of education. Whewell's *Philosophy* represented a concerted effort to evaluate the place of scientific knowledge in human experience. His descriptive view of mathematics firmly located the subject as one of the sciences. As such, it brought questions about the nature of mathematical knowledge out of a specialized realm and into a broader discussion of the nature of scientific knowledge and its place in intellectual culture.

In her recent book, *Science and Culture*, Susan Cannon fixed on a unitary view of truth as one of the salient aspects of the English

---

[13] Ibid., pp. 148–9.

[14] William Whewell, "Of Some Objections Which have Been Made to the Doctrines Stated in the Previous Chapter," *The Philosophy of the Inductive Sciences*, 2nd ed. rev., 2 vols. (London, 1847; reprint ed., London: Frank Cass & Co. Ltd., 1967) 1:101–11.

approach which most influenced Victorian scientific style. The classic statement of this perspective is Herschel's: "The grand and indeed only character of truth, is its capability of enduring the test of universal experience, and coming unchanged out of every possible form of *fair* discussion."[15] This view that truth was singular, fixed and bound to emerge from any honest and careful attempt to discover it, was reflected in the intellectual lives of British scientific polymaths who moved unself-consciously between seemingly disparate scientific and humanistic fields.[16] It was evident in resistance to dividing the meetings of the BAAS into sections.[17] It was also implicit in the faculty approach to education which was built on the assumption that the same intellectual tools were relevant to all subjects. Whewell's defense of mathematics at Cambridge was in an important sense a defense of a particularly powerful unitary view of truth which included mathematics as an essential part of the closely woven tapestry of human knowledge.

Whewell was a nativist. This means that he thought that the basic truths which were being discovered and explored in the sciences were innate. In ignorant people, ignorant either because uneducated or living before some particular truths were known, these ideas were present but not recognized or known. Learning essentially involved identifying the idea or ideas intrinsic to a given situation. For Whewell, the essence of inductive reasoning lay in appropriately applying innate ideas to the external world. Whewell's nativism set him apart from most of his compatriots who were basically empiricists. They would maintain that in induction ideas were somehow drawn from experience. However, Whewell's specific claim that the interpretation of mathematical symbols was the basis for their truth was widely accepted. More typically British thinkers also developed views of mathematics which emphasized its similarities to other kinds of knowledge. For these empiricists, as for the nativistic Whewell, the truth of mathematics continued to rest on its descriptive power. The empirical approach can be seen in the works of John Herschel, who, like Whewell, was initially visible as a part of the Analytical Society.

---

[15] Herschel, *Preliminary Discourse*, (1830) p. 10.
[16] Schweber (1981) emphasizes this characteristic of British scientists.
[17] Morrell and Thackray (1981), p. 452.

Herschel briefly expounded an empirical philosophy in his *Preliminary Discourse* of 1831, and subsequently refined it in a number of shorter articles. In contrast to Whewell's nativism, Herschel emphasized that the mind somehow drew its ideas out of experienced situations. He defended a variant of Locke's classic theory of the *tabula rasa* against Whewell's quasi-Kantian insistence on preprogrammed forms of understanding.

Herschel's emphasis on the central importance of experience, and the concomitant rejection of Whewell's innate Fundamental Ideas, necessitated that he develop a different interpretation of the nature of mathematics. Whereas Whewell maintained that mathematical development entailed unraveling these complex intellectual constructs, Herschel interpreted it as the description of a concrete, objective subject matter. Space, the subject of geometry, was an external reality; magnitude, the subject of arithmetic, was also an objective fact. To quote from one of Herschel's later essays:

The truths of geometry *exist* and are verified in every part of space, as the statue in the marble. They may depend on the thinking mind for their conception and discovery, but they cannot be contradictory to that which forms their subject-matter, and in which they are realized, in every place and at every instant of time.[18]

Herschel claimed that the only distinction between empirical sciences and mathematics was one of degree. In the sciences generally considered inductive, the passage from experience to generalities was difficult enough that much of the historical effort to generate the sciences had been expended on it. Mathematical experiences, on the other hand, were so universal, their relations so simple, that the basic axioms had been easily found. Since the major portions of the energy devoted to these sciences was directed at their deductive aspects, their inductive roots were easily overlooked. Thus, Herschel wrote,

The only difference between these [the axioms of geometry] and axioms obtained from extensive induction is this, that, in raising the

---

[18] [John Herschel], "Review of Whewell's *History of the Inductive Sciences* (1837) and *Philosophy of the Inductive Sciences* (1840)," *The Quarterly Review*, 68 (June and September, 1841):207.

axioms of geometry, the instances offer themselves spontaneously, and without the trouble of search, and are few and simple; in raising those of nature, they are infinitely numerous, complicated and remote; so that the most diligent research and the utmost acuteness are required to unravel their web, and place their meaning in evidence.[19]

Within this kind of an empirical philosophy, the easy accessibility of mathematical truths was so striking that, although theoretically they were gained in the same way as scientific truths, practically pursuing them would be very different. At one point Herschel even proposed that the concepts of number and space might somehow arise from prenatal or ancestral experience. For this reason, in some of his early writings, Herschel questioned the value of studying mathematics as part of a broadly based education in much the same way that Hamilton had done. In the long run, however, Herschel firmly asserted the similarities between mathematics and the sciences and staunchly defended a quasi-Whewellian descriptive view of the subject against those who might argue against that approach. The crucial issue which solidified Herschel's position on this point, an issue which continued to be closely implicated in mathematical discussions throughout the rest of the century, concerned the nature of truth.

In the debate between Whewell and Hamilton, mathematics was considered primarily as a vehicle for teaching men to reason. From this perspective it was critically important that the kind of thinking that went on in mathematics be seen as similar to that which was basic to other forms of thought. If one accepted the Scottish definitional view which made mathematical thinking different from other subjects, it quickly lost its educational validity. From a nativist point of view, mathematics provided the purest example of effective reasoning, but to an empiricist its claims for centrality were more tenuous. It was difficult to argue that the best way to learn to reason empirically was to study mathematics. However, it is too narrow to picture the nineteenth-century English ideal of the liberal education as simply focused on the development of reasoning or any other

---

[19] Herschel, *Preliminary Discourse*, (1830), p. 96.

intellectual powers. The descriptive view of mathematics Whewell advocated against the Scots was also important on another level; it carried important implications for an entirely different discussion about the nature of humanly accessible truth.

Mathematics and Truth: William Whewell and John Herschel

The most essential goal of the liberal education at both Cambridge and Oxford was humanistic, to educe and cultivate the purest core of humanity in the students. The shift in emphasis from cultivating manners to developing the intellect, as mentioned previously, reflected important changes in prevailing notions of the nature of human life. There were further distinctions within the nineteenth-century intellectual vision. At Oxford, the humanistic goal was pursued through the study of the classics. It was hoped that by acquainting themselves with the greatest minds and lives of past civilization, students would find the way to develop their own human potential to the fullest. Leaving aside questions about the value of the studying the ancients as opposed to the moderns, etc., the rationale behind this curricular design seems fairly straightforward. At Cambridge, however, the same humanistic goal was to be attained through mathematical study. This choice requires somewhat more elaboration to be comprehensible. The arguments in defense of mathematics reveal a second important implication of the mid-century descriptive view.

At some level, interpreting the mathematical education at Cambridge as a means to a liberal education involved seeing the highest aspiration of human life as modeled on the scientist. This point found striking expression in Herschel's *Preliminary Discourse* where he advocated a general scientific education. Even for those who were unable to actively contribute to its progress, a basic understanding of science would "form, as it were, a link between ourselves and the best and noblest benefactors of our species, with whom we hold communion in thoughts and participate in discoveries which have raised them above their fellow mortals, and brought them nearer to their Creator."[20]

---

[20] Ibid., pp. 16–17.

This quotation from Herschel points to a second aspect of the English view of truth Cannon identified as characteristic. To the Victorians, science was not only an integral part of a unified truth complex, it served as the norm of truth.[21] Scientists were the leaders in the human search for truth. It was they who in their researches were coming close to the divine. The strength of this kind of argument is attested to by the strength of British natural theology, for example in the Bridgewater treatises which established religious truths on a scientific basis. In mid-century England success in finding truth in any area tended to be judged in terms of how close it came to the perceived standard of scientific truth.

This attitude focused attention on the question of the ultimate validity of science: how well do we know scientific truths? In his philosophical work, Whewell had focused extensively on this basic epistemological problem phrased as whether man could come to an absolute knowledge of the laws of nature. Again, though his nativism made him somewhat idiosyncratic in empirically oriented England, his views molded much of his compatriots' subsequent discussion.

In order to address the issue of whether or how humans could know truth through science, Whewell drew a sharp distinction between contingent truths, which were merely summaries of observed phenomena, and necessary truths, which captured the essence of reality. Contingent truths were observed facts about phenomena which "for aught we can see . . . might have been otherwise."[22] Necessary truths, on the other hand, were those whose opposites were inconceivable.

Whewell illustrated the distinction between necessary and contingent truth with Kepler's and Newton's work on elliptical planetary orbits. When Kepler concluded that planets rotated in elliptical orbits, it was imaginable to him that they might have followed a different path; it was only his observational data which led him to this conclusion. In Whewell's mind, this kind of understanding of planetary orbits was in marked contrast to Newton's. Because he under-

---

[21] Cannon, "Science as the Norm of Truth," *Science in Culture*, (1978), pp. 1–28.
[22] W[illiam] Whewell, "On the Fundamental Antithesis of Philosophy" (Read Feb. 5, 1844), *Transactions of the Cambridge Philosophical Society*, 8 (1849):170.

stood the nature of force, Newton knew not only that the orbits were elliptical but also that they could be nothing *but* elliptical. In his clearly conceived concept of force, Newton's mind was in essential harmony with the external world. Whewell maintained that whereas Kepler had developed only a *contingent* description, Newton knew the form of planetary orbits *necessarily*.

Whewell's category of necessary truth was critically important for the assurance that man really could come to know his world. This assurance in turn supported his basically conservative outlook in which there were certain immutable truths about God, man and society which the educated elite, of which he was a member, understood and passed down from generation to generation. His category of necessary truth was not only relevant to abstract epistemological arguments, but in the 1840s, buttressed a much broader and highly conservative political, social and theological outlook.

Whewell used the example of mathematics to establish the reality of necessary truth, and to demonstrate that human minds could grasp it. In geometry he pointed out, one not only knew that the sum of the angles of a triangle was 180°, it was impossible to imagine that it could be otherwise. When a theorem like this one was fully understood, it was impossible to clearly conceive of its being contradicted: such a conception would inevitably violate some other indubitable piece of knowledge like the nature of the straight line, the right angle or the triangle.

For Whewell, this kind of certainty was possible because in mathematics there was no distinction between objective reality and subjective knowledge; the human mind was completely in tune with external mathematical fact. Thus he wrote,

> If any person does not fully apprehend, at first, the different kinds of truth thus pointed out [contingent and necessary], let him study, to some extent, those sciences which have necessary truth for their subject, as geometry [the study of space], or the properties of numbers [arithmetic], so as to obtain a familiar acquaintance with such truth; and he will then hardly fail to see how different the evidence of the propositions which occur in these sciences, is from the evidence of the facts which are merely learnt from experience.[23]

---

[23] Whewell, *Philosophy*, (1847), vol. 1, p. 58.

Whewell developed his interpretation of mathematics within the context of an anti-empirical, intuitionist view of science which was in many ways far from the mainstream of English thought. His ideas were very powerful in mid-century English science, however, not merely because of his position as master of Trinity College, but because they were an expression of attitudes important to many of the now aging members of the Analytical Society. The turbulent '30s and '40s had raised pressing questions for which the intellectually attainable necessary truth Whewell promised on the basis of mathematics provided important answers.

Whewell's position found striking support in Herschel's "review" of his work, which appeared in 1841. This review is interesting not only because it contains an empirical reinterpretation of many of Whewell's attitudes, but also because it points to some of the unstated implications of Whewell's emphasis on necessary truth. Herschel's earlier work would not have supported Whewell's view. The development of his thought around this point is indicative of the wide range of changing concerns which were implicated in English discussions of geometry throughout the rest of the century.

In his 1830 *Preliminary Discourse,* Herschel had written that the mission of the natural philosopher lay in

> the endeavor to discover, as far as our faculties will permit, what *are* these primary qualities originally and unalterably impressed on matter, and to discover the *spirit* of the laws of nature, which includes groups and classes of relations and facts from the *letter* which . . . is presented to us by single phenomena: or if, after all, this should prove impossible; if such a step be beyond our faculties; and the essential qualities of material agents be really *occult,* or incapable of being expressed in any form intelligible to our understandings, at least to approach as near to their comprehension as the nature of the case will allow; and devise such forms of words as shall include and *represent* the greatest possible multitude and variety of phenomena.[24]

This passage suggests that in his early thinking, Herschel did not consider science valuable primarily as a truth-seeking activity; the

---

[24] Herschel, *Preliminary Discourse,* (1830), p. 39.

unambiguous establishment of scientific truth seems to have been almost peripheral for him. Rather than emphasize that scientific practitioners were guardians of truth, his youthful reverence for them rested on what he saw as the ennobling potential of scientific pursuits. He focused on the process of scientific study more than on the perfection of the results attained. To the young Herschel, pursuing science revealed "our strength and innate dignity"; it filled us, "as from an inward spring with a sense of nobleness and power."[25] He felt that scientific study would enable the educated to "participate in discoveries":[26] the truth of these discoveries *per se* was not the central issue.

Herschel's youthful emphasis on the intellectually and spiritually liberating character of scientific study changed over the years. By 1841, when he critically reviewed Whewell's *Philosophy,* his emphasis on process was giving way to a focus on truth. In this review, Herschel unambiguously repudiated his earlier agnosticism about man's intellectual capacities. He shifted his focus away from the methods of generating scientific knowledge to the truth of the knowledge so gained. In his response to Whewell it was scientific truth which was divine, not the method by which it was attained.

The change in Herschel's emphasis is reflected in the summary treatment he offered of the major distinction between Whewell's nativist philosophy and his empirical one. He characterized the difference as follows:

Do we apply to the objects of our reasoning, ideas of which we have a perception, and propositions of which we have a conviction antecedent to experience—(and which may therefore be regarded as impressed on our intellectual nature by the Author of our being) . . .? Or do we simply distribute all the phenomena of the world around us, and of our own minds, into groups, according to the analogies of the impressions they make on our perceptive faculties, whether bodily or mental—(the perception of such analogies being itself one of the primordial faculties of our minds;) . . . ?[27]

---

[25] Ibid., p. 16. The full quotation is on p. 14.
[26] Ibid., p. 17. The full quotation is on p. 27.
[27] Herschel, "Review," (1841), p. 181.

Using this methodological distinction to differentiate between nativism and empiricism, Herschel indicated that Whewell's views fell primarily into the first category, and his into the second.

As soon as he had clarified this dichotomy, however, Herschel suggested that it was not fundamentally important. "And after all," he wrote,

> it seems far from certain that this opposition of views is anything more than apparent; for among the infinite analogies which may exist among natural things, it may very well be admitted that those only are designed, in the original constitution of our minds, to strike us with permanent force, to embody around them the greatest masses of thought and interest, to become elaborated into general propositions, and finally to work their way to universal reception, and attain to all the recognizable characters of truth, which are really dependent on the intimate nature of things as that nature is known to their Creator, and which have relation to their essential qualities and conditions as impressed on them by Him; so that the power bestowed on the mind of seizing on those primordial analogies, and its impulse to generalize the propositions which their consideration suggests, . . . are equivalent to its endowment with a direct recognition of fundamental ideas and relations not derived from experience, and the evolution from those ideas of necessary truths equally independent of experience, in the other.[28]

Although their interpretations of method might differ, Herschel felt that little essentially separated him from Whewell as long as they agreed on the solidity of scientific truth. This is because a central issue for both of them in considering science was to firmly establish the validity of the knowledge of God. By admitting the possibility of divine aid in man's empirical search for knowledge, Herschel allowed for a source of information which was not totally reliant on our imperfect senses. Thus, even within his empirical approach to science, it was possible to gain a solid knowledge of God's truth. "And," he continued,

> perhaps, with this explanation both parties ought to rest content— satisfied that, on either view of the subject, the mind of man is rep-

---

[28] Ibid., p. 182.

resented as in harmony with universal nature; that we are consequently capable of attaining to real knowledge; and that the design and intelligence which we trace throughout creation is no visionary conception, but a truth as certain as the existence of that creation itself.[29]

Clearly, to the mature Herschel, the nature of scientific truth was a pressing issue which was closely tied to theological argument. For this reason, when he responded to Whewell's *Philosophy* he took pains to make it clear that he believed man could attain real knowledge through science, and centrally, mathematics. Viewing mathematics from this perspective, however, necessitated that mathematical systems be interpreted as carefully constructed descriptions of an independently existing subject matter. Herschel was specific about the impossibility of attributing significant truth to a mathematical statement on any except this kind of descriptive basis.

It may, however, be alleged, that one criterion of abstract truth remains unconsidered—its direct recognition *in the abstract* without mental reference to *any* particular case, to *any* example, to *any* experience. How truth may or may not impress conviction in other minds, it is doubtless presumptuous to assert, for which reason we have dwelt only on the received *tests* of truth, as conveyed from mind to mind by the intervention of language. If there be those who can persuade themselves that they are yielding a rational assent to the terms of an abstract proposition on the mere jingle of its sound in their ears, while refusing to test it by calling up in their minds those images with their attributes, which experience has inseparably associated with its words, they have certainly a very different notion of logical evidence from our own.[30]

As this passage suggests, Herschel's mid-century focus on clearly established mathematical truth was sharp enough that it invalidated any kind of formal approach to the subject. To be significant, the truth of mathematical statements had to lie in the subject matter being described; it could not be confined to empty manipulations of terms and figures which were merely subjectively defined.

---

[29] Ibid.
[30] Ibid., pp. 221–22.

Thus, by the middle of the century, although apparently poles apart philosophically, Herschel and Whewell seem to have been very close together in the essential outlines of their views of mathematics. Both of them insisted that significant mathematics was essentially descriptive and argued vehemently against suggestions that it might be otherwise conceived.

The institutional backdrop to this point of agreement between Herschel and Whewell suggests that it is not coincidental. Both men were educated at Cambridge where the mathematics they learned was embedded in the liberal education. Their subsequent interest in the nature of mathematics continued to be bound up in the development of this institutional framework. On an important level their philosophies of mathematics were defenses of their *alma mater* and the education it offered.

For the historian, recognizing the educational context in which they forged their views of mathematics is helpful because it points to the tangle of changing intellectual and other issues which were implicated in various formulations of the nature of the subject. At the same time, however, it is tempting to dismiss the importance of their ideas since they might appear to be idiosyncratically parochial. To view Whewell and Herschel in this way, however, would be to miss the central importance of Cambridge mathematics to the English view of the subject. Perhaps the best way to illustrate how widespread and powerful the Cambridge-based vision was, is to look at the work of John Stuart Mill.

Mathematics and Science: John Stuart Mill

For all intents and purposes, John Stuart Mill was not enmeshed in the traditions and developments at Cambridge. His entire education, which began when he was an infant, was carefully constructed on utilitarian principles. His father's goal in bringing him up was to form a new kind of person who would be free of all the irrational impulses and values which were enmeshed in the ambient English culture. Even after the younger Mill's nervous breakdown, which precipitated a break with much of what his father had taught him, Mill stood firmly outside of the established intellectual centers in

England and developed his ideas in self-conscious opposition to those being supported at Cambridge and Oxford.

This is certainly the case with Mill's philosophy of science. Mill's major work, *A System of Logic*, first appeared three years after Whewell's *Philosophy*. In this work, Mill tried to present a systematic empirical alternative to the nativist approach Whewell had espoused. He was strongly opposed to the conservative overtones which lurked in Whewell's relentless emphasis on the primacy of esoteric immutable principles for understanding the world. Instead, Mill depicted an epistemologically democratic universe, which was known through common experience, not through rare insights carefully guarded and passed down through the generations.

Mill's major target in this philosophical battle was Whewell's nativist philosophy. In his *Autobiography*, Mill is charmingly candid about the importance of Whewell as an antagonist for the composition of his *Logic*.

> During the re-writing of the Logic, Dr. Whewell's Philosophy of the Inductive Sciences made its appearance; a circumstance fortunate for me, as it gave me what I greatly desired, a full treatment of the subject by an antagonist, and enabled me to present my ideas with greater clearness and emphasis . . . in defending them against definite objections, or confronting them distinctly with an opposite theory.[31]

Most of the supporting evidence he used to argue his case came from Whewell's *History of the Inductive Sciences*. In and of itself this indicates the tremendous impact of Whewell's work on mid-century English views of the nature and development of science.

By selecting Whewell as his opponent, however, Mill placed himself in the position of arguing on someone else's ground. At least in mathematics his reliance on his antagonist to sharply define the basic points at issue had a significant effect on his position. Mill was caught in the fray and became like his enemy. Despite his attempts to develop an empiricism which would confute Whewell's emphasis

---

[31] *Autobiography of John Stuart Mill*, with a Preface by John Jacob Coss, (New York: Columbia University Press, 1924), p. 156.

on the critical importance of innate ideas, Mill ultimately defined mathematics in essentially the same way.

Mill's major thesis in the *Logic* was that scientific reasoning was not about ideas in the mind, but rather about objective facts of nature. The relationships science recognized and established were not connections among subjective impressions of things but rather connections among the things themselves. Thus, objective rather than subjective reality was the primary locus of scientific truth.

This orientation led Mill to reconsider the role and nature of definitions in mathematics and science. With his nativistic focus on Fundamental Ideas, Whewell had emphasised the central role of subjectively formulated definitions for the construction of true mathematical and scientific theories. Mill's empirical emphasis on objective truth conflicted directly with this approach. Basically, Mill claimed, definitions are verbal. They do not relate directly to objective matters of fact, but rather to subjective meanings of words. Thus they are not, strictly speaking, true or false but only more or less acceptable as ways of discussing objective reality. Thus, if in mathematics the entire system rested on arguments built upon abstract definitions, it would be an essentially subjective system. As such it might have its uses, but it would not be able to claim scientific veracity. It would seem, then, that Mill's approach would lead him to a position very close to Hamilton's.

In the event, though, Mill did not go this far. He tempered the starkly subjective approach to definitions in response to the nativist arguments of Whewell's *Philosophy*. Whewell had there argued so powerfully that definitions played a central role in what Mill saw as truly objective scientific arguments that Mill complicated his interpretation of their status. He had to account for the central place that definitions played in indubitably true scientific theories: he had to interpret the importance of the concept of force in Newton's physics, for example.

Mill approached this problem by including considerations of existence in his treatment of definitions. Although definitions are essentially verbal, he argued, truly *scientific* definitions are accompanied by implied postulates of existence. These definitions are not merely subjective; they point beyond themselves to real, objective entities.

Embedding an existence postulate in scientific definitions enabled Mill to locate the truth of scientific theories in experienced, external reality rather than in the conceiving mind. Demonstrative truths which appeared to flow from subjectively constructed definitions were, in fact, falling out of the existence postulates implied in the definitions. They were the bridge between subjective and objective experience on which scientific truth rested. "There is a real distinction," Mill wrote,

> between definitions of names, and what are erroneously called definitions of things; but it is, that the latter, along with the meaning of a name, covertly asserts a matter of fact. This covert assertion is not a definition, but a postulate. . . . It affirms the real existence of Things possessing the combination of attributes set forth in the definition; and this, if true, may be foundation sufficient on which to build a whole fabric of scientific truth.[32]

Thus, for Mill, scientific statements were affirmations of the existence of their referents as well as being descriptions of the relations among them.

Mill's claim that truly scientific definitions implied objective existence led him to consider the case of mathematics, because here it was not clear in what sense a definition could be claimed to assert existence. Whewell, Hamilton, Herschel and Mill all would have agreed that there are no such things as real, objective objects corresponding to mathematical definitions; a true, mathematical circle could never be found. Nevertheless, Mill felt that true mathematical conclusions seem to be deduced from definitions. Here he was faced with either embracing Hamilton's empty interpretation of mathematical truth and at the same time making a distinction between mathematical and scientific truth or asserting the ultimate unity of truth by adopting some kind of descriptive mathematics. In this case, Mill chose Whewell's side. Mathematical definitions, he claimed, are like scientific ones because they contain implicit exis-

---

[32] John Stuart Mill, *A System of Logic, Ratiocinative and Inductive,* (New York: Harper and Brothers, 1846), p. 99.

tence postulates. Thus, the mathematical definition of a circle points to something real which exists outside of the definition. It is the fact that the mind can form a notion of these circles that is the hidden postulate of the definition on which the mathematical argument is based.

Whewell and Mill did not agree about the source of these mathematical ideas. For Mill, the empiricist, mathematical ideas were somehow constructed inductively from experience; for Whewell, the nativist, they were innate. They did agree, however, that the truth of mathematical demonstrations rested on these ideas. The validity of theorems about circles does not flow from abstract, verbal terms of the definition itself even though the argument might be constructed deductively from those terms. Ultimately the mathematical truth that a circle has 360° rests on the idea of a circle and that of degrees, not on the demonstration of the theorem.

In this, then, which was the basic point of contention between Whewell and the Scottish school, Mill fell on the side of Whewell. His view, he noted,

> leaves the conclusion that our reasonings are grounded upon the matters of fact postulated in definitions, and not upon the definitions themselves, entirely unaffected; and accordingly I am able to appeal in confirmation of this conclusion, to the authority of Mr. Whewell, in his recent treatise on *The Philosophy of the Inductive Sciences*. On the nature of demonstrative truth, Mr. Whewell's opinions are greatly at variance with mine, but on the particular point in question it gives me great pleasure to observe, that there is a complete agreement between us.[33]

As the last sentence indicates, Mill rejected the view of absolute necessary truth which Whewell had constructed on his descriptive mathematics, but he could not resist agreeing with that view itself.

The rarity of issues on which Mill and Whewell agreed served to emphasize the strength of the descriptive view of mathematics in England in the middle of the nineteenth century. Clearly it was a view which both bolstered and was bolstered by the mathematical

---

[33] Ibid., pp. 102–03.

education at Cambridge. The case of Mill, however, who was not directly connected to this institution in any way, indicates the cogency of the intellectual as well as the institutional constructs which supported it.

Thus, at the end of the first half of the century a variety of important issues focused interest on the nature of mathematics. Among them was concern about the relation of mathematical reasoning to reasoning in other fields. Another concerned the nature of humanly accessible truth. Discussion of these questions was often closely tied to the education at Cambridge, but both of them had implications which transcended this context. A descriptive view of the nature of mathematics was central to resolving questions about the nature of mathematical reasoning and truth in such a way as to preserve its position at the center of the Cambridge curriculum. The influence of this approach extended far beyond Cambridge, however. It affected virtually all English mathematical discussion well into the century.

ALGEBRA VS. GEOMETRY WITHIN A DESCRIPTIVE VIEW OF MATHEMATICS

The philosophically elaborated, descriptive view of mathematics which emerged in the middle of the century carried with it some specific implications for what kinds of study were most valid. It is significant that the mathematical examples Whewell, Mill or Herschel offered in support of their arguments about the nature of mathematical knowledge were consistently geometrical as opposed to analytic. Geometrical arguments are more clearly descriptive than analytical ones. To argue that a proof involving circles requires a conception of space is much easier than that an analytical demonstration involving a and b requires an understanding of number. In an important sense, then, the argument about the legitimacy of mathematics at Cambridge implied a judgment about the kinds of mathematics which somehow were most natural and legitimate. This judgment had specific implications for the kinds of mathematics which were pursued, as well as the way questions were approached.

The nature of these implications was explicitly negotiated in the mid-century discussions about the Cambridge Tripos.

Before 1848, the Tripos was an undifferentiated six-day examination. In the reform of 1848 it was lengthened to eight days and divided into two parts. The first three days were designed to cover the material essential for anyone to receive an ordinary degree. Only after he had completed the first part of the examination could a student sit for the more advanced, second part of the examination. His performance on this second part determined whether he would receive honors. Until 1851 students had to receive mathematical honors before they were allowed to compete for honors on the Classical Tripos. After that date, when the Moral Sciences Tripos and Natural Sciences Tripos were added, students could attempt to receive honors on any Tripos after taking only the first part of the Mathematical Tripos. Thus, until the end of the century, the first part of the Mathematical Tripos remained the solid core of the education of any Cambridge graduate.[34] This meant that when they fixed the form of the first part of the Tripos, the reformers were deciding on those subjects which would constitute the backbone of the Cambridge education. The issues surrounding this decision illustrate the kinds of implications attendant on embedding mathematics into the Cambridge liberal education and more generally into English intellectual culture as a whole.

Cambridge educators did not treat the first days of the Tripos as merely more elementary than the second part. A Board of Mathematical Studies was formed which laid out strict guidelines for the examination which went significantly beyond considerations of simplicity. According to the 1849 "Report of the Board of Mathematical Studies," the first part of the examination was to test

> the portions of Euclid usually read [first four books]; Arithmetic; parts of Algebra, embracing the Binomial Theorem and the Principles of Logarithms; Plane Trigonometry, so far as to include the solution of Triangles; Conic Sections, treated geometrically; the elementary parts of Statics and Dynamics, treated without the Differential

---

[34] A fuller discussion of the Tripos is to be found in Harvey W. Becher, "William Whewell and Cambridge Mathematics," *Historical Studies in the Physical Sciences*, 11 (1980):1–48, or D. A. Winstanley, *Early Victorian Cambridge*, (Cambridge: Cambridge University Press, 1940).

Calculus; the First three Sections of Newton, the Propositions to be proved in Newton's manner; the elementary parts of Hydrostatics, without the Differential Calculus; the simpler propositions of Optics, treated geometrically; the parts of Astronomy required for the explanation of the more simple phenomena, without calculation.[35]

One of the most striking aspects of this list is its emphasis on what would now be seen as applied subjects. It is applied mathematics and theoretical physics which most clearly fit the Cambridge criteria of educational value.

Within this over-arching characterization, however, is another, finer one which carried the descriptive focus deep into the purely mathematical parts of the program. The list goes beyond the narrow goal of specifying what subjects should be included, to establish strict criteria for how they should be approached. Thus dynamics is to be taught "without the Differential Calculus," the preliminary sections of the *Principia* to be taught with "the propositions proved in Newton's manner."

Each of these specifications reflects a descriptive view of mathematics. They require that diverse mathematical subjects be taught by geometrical methods, even when more efficient analytical ones were readily available. Thus, in an important sense, the 1848 reform marked a renewed emphasis on geometry at the expense of analysis—a conscious return to the intuitive emphasis in mathematics which had been displaced three decades before.

This emphasis on geometry at the expense of analysis was the culmination of a tendency which can be seen growing in intensity in the work of maturing members of the Analytical Society during the 1830s and 1840s. In 1835, when he wrote "Thoughts on the Study of Mathematics," Whewell had advocated mathematical study in general as the best means for the education of young men. In the next decade, he modified this early position and began to advocate the study of geometry as opposed to analysis. In 1845 he wrote,

mere analytical reasoning is a bad discipline of the intellect, on account of the way in which it puts out of sight the subject matter of

---

[35] Cambridge University Commission, "Appendix to Evidence from the University," (1852), p. 232.

the reasoning; on the right apprehension of which, with its peculiar character and attributes, all good reasoning, on all other subjects, must depend.

It is easy to shew by examples . . . that analytical modes of treating . . . those subjects have, in fact, put out of sight the peculiarities of the conceptions which belong to each subject, and have merged all their special trains of reasoning into undistinguishing symbolical generalizations.[36]

Whewell here criticized analysis because the power of its symbology obscured the focus on a particular subject matter which was essential to valid science. For this reason, geometry, which *required* that one concentrate directly on the known properties of space in order to solve particular problems, served as significantly better training.

As he sharpened his philosophical position, Whewell became more specific about what kind of mathematics properly served his educational goals. During the same period, between the *Preliminary Discourse* of 1830 and an 1845 presidential report to the British Association for the Advancement of Science, Herschel's ideas developed along a parallel route. The young Herschel would have rejected Whewell's descriptive emphasis in mathematical study. In the *Preliminary Discourse,* he had asserted that scientific connections were recognized by drawing analogies among far-flung disciplines. Generality was the mark of scientific power, and analogies which could be found to hold among many disparate phenomena bore with them the characteristics of *vera causa* and the stamp of truth. In this view, the flexible generality of analysis was closer to creative scientific thinking than geometry, and therefore more valuable.

Herschel's emphasis changed markedly over the years, however. His later views are contained in his 1845 presidential address to the BAAS meetings held at Cambridge. It is clear from this speech that recent scientific speculations had troubled him. On the one hand, he was disturbed by the evolutionary speculations contained in the then anonymous *Vestiges of Creation.* From Herschel's standpoint the arguments in this work emphasized "loose and vague" descriptive laws over carefully analysed explanatory causes. This approach

---

[36] William Whewell, *Of a Liberal Education in General,* (London: John W. Parker, 1845), p. 45.

was a betrayal of the ultimately truth-seeking enterprise Herschel felt science to be. Relatedly, he was critical of the "nebulous hypothesis" which had been first proposed by LaPlace and developed by Comte. Here again, the impulse to generalize and to "[throw] overboard as troublesome" all manner of essential considerations had led Comte and others to exaggerated claims for the theory. Herschel commented,

> I really should consider some apology needed for even mentioning an argument of the kind to such a meeting, were it not that this very reasoning . . . has been eagerly received among us* as the revelation of a profound analysis. When such is the case, it is surely time to throw in a word of warning, and to reiterate our recommendation of an early initiation into mathematics, and the cherishing of a mathematical habit of thought, as the safeguard of all philosophy.
>
> *Mill. Logic, ii.28 — Also, 'Vestiges of the Creation,' p. 17.[37]

As this passage hints, Herschel saw the study of mathematics as the proper palliative for such wrong-headed and empty speculation. Just what he meant by "a mathematical habit of thought" is suggested by the paean to the Cambridge mathematical education with which he opened his address.

> It has been, and I trust it ever will continue to be, the pride and boast of this University to maintain, at a conspicuously high level, that sound and thoughtful and sobering discipline of mind which mathematical studies imply. Independent of the power which such studies confer as instruments of investigation, there never was a period in the history of science in which their moral influence, if I may so term it, was more needed, as a corrective to that propensity which is beginning to prevail widely, and, I fear, balefully, over large departments of our philosophy, the propensity to crude and over-hasty generalization. To all such propensities the steady concentration of thought, and its fixation on the clear and the definite which a long and stern mathematical discipline imparts, is the best, and, indeed, the only proper antagonist.[38]

---

[37] Sir John F. W. Herschel, "Presidential Address," *Report of the Fifteenth Meeting of the BAAS Held at Cambridge in June 1845*, (London: John Murray, 1846), p. xxxix.
[38] Ibid., pp. xxvii–xxviii.

Herschel's concern that scientific theorizing be strictly con-
strained by right thinking convinced him that mathematical study
must be approached cautiously. The "long and stern mathematical
discipline" he advocated in 1845 was a far cry from the exuberant
open-ended analytic development he and his friends had ushered in
thirty years earlier. The troublesome directions in which this free
kind of thinking seemed to have gone led him to see the wisdom of
a mathematical training which emphasized "the steady concentra-
tion of thought, and its fixation on the clear and definite" over general reasoning.[39] Within the truth-oriented discussions of the mid-
century it was the conceptual clarity of geometrical understanding
that was the critical message of mathematical study.

This concern with maintaining the purity of mathematical truth
was not only relevant to the design of elementary curricula. The
views of knowledge and mathematics Herschel and Whewell
defended were not offered merely as palliatives for young minds.
They were concerns which permeated English mathematical thinking all the way to its highest levels.

Since neither Herschel nor Whewell was actively engaged in
mathematical research by the middle of the century, these men are
not helpful for illustrating the far-reaching implications of their
mature view of mathematics. The work of their slightly younger
compatriot, Augustus De Morgan, however, illustrates the internal
mathematical implications of their descriptive approach. De Morgan can be regarded as a satellite of the Analytical Society. Somewhat isolated and fiercely independent, he nonetheless corresponded regularly with many of these men and tried to promote the
study of mathematics from the somewhat academically provincial
outpost of University College.

From the publication of his first book, a translation of Bourdon's
Elements of Algebra, to the end of his life, De Morgan was concerned
with the nature of algebra.[40] During the 1830s, most strikingly in an

---

[39] A fuller discussion of the philosophical issues surrounding this report is in Silvan Schweber,
"Auguste Comte and the Nebulous Hypothesis," forthcoming.

[40] For a more detailed treatment of De Morgan's development see Helena M. Pycior, "The
Three Stages of Augustus De Morgan's Algebraic Work," *Isis,* 74 (1983):211–22; Joan L.
Richards, "Augustus De Morgan, the History of Mathematics, and the Foundations of Alge-

1835 "Review" of Peacock's *Algebra,* De Morgan entertained the possibility that algebra might entail simply an internally consistent symbolic development which was more or less susceptible to any particular interpretation. From the very beginning, though, De Morgan was clearly somewhat uncomfortable with this view, noting he paused for five years before writing it because "[at] first sight it seemed to us like symbols bewitched, and running about the world in search of meaning." The loss of meaning it entailed seemed to severely undercut the validity of analytical researches. However, De Morgan endorsed it in 1835, at least until a better one could be developed.[41]

In his later writings, De Morgan returned to this point and clarified his emphasis. Although he did not deny that it was possible to develop an abstract symbolic algebra without specifying its interpretation, he did make it clear that this kind of study was essentially valueless. Legitimate mathematical study did not end with formal consistency. In the paper "On the Foundation of Algebra," delivered to the Cambridge Philosophical Society on December 9, 1839, De Morgan distinguished between technical algebra—"the art of using symbols under regulations which, when this part of the subject is considered independently of the other, are prescribed as the definitions of the symbols"—from the logical algebra—"the science which investigates the method of giving meaning to the primary symbols, and of interpreting all subsequent symbolic results."[42] Having made this distinction, De Morgan proceeded to elaborate on the complete algebraic process which involved both of these aspects. He wrote,

Thus a symbol is *defined* when such rules are laid down for its use as

---

bra", *Isis,* 78 (1987):7–30. More expanded considerations of algebra in England are to be found in Joan L. Richards, "The Art and Science of British Algebra: A Study in the Perception of Mathematical Truth," *Historia Mathematica,* 7 (1980):343–65; Helena M. Pycior, "George Peacock and the British Origins of Symbolical Algebra," *Historia Mathematica,* 8 (1981):23–45; Helena M. Pycior, "Early Criticisms of the Symbolical Approach to Algebra," *Historia Mathematica,* 9 (1982):413–40.

[41] Augustus De Morgan, review of George Peacock, *A Treatise on Algebra,* in *Quarterly Journal of Education,* 9 (1835):311.

[42] Augustus De Morgan, "On the Foundation of Algebra," (Read Dec. 9, 1839), *Transactions of the Cambridge Philosophical Society,* 7 (1842)173.

will enable us to accept or reject any proposed transformation of it, or by means of it. A simple symbol is *explained* when such a meaning is given to it as will enable us to accept or reject the application of its definition, as a consequence of that meaning: and a compound symbol is interpreted, when, having occurred as a result of explained elements, used under prescribed definitions, a necessary meaning can be given to it; the necessity arising from the tacit supposition that the compound symbol, considered as a new simple one, must still be subject to the prescribed definitions, when it subsequently comes in contact with other symbols.[43]

This description of the process of algebra moves in much the same way that Mill had done, from highly abstract definitions of terms to a focus on the objects they describe. Although terms are "defined" implicitly by their position within the forms, the process of algebraic investigation does not stop here. It continues through "explanation," the assignation of meaning to the simple symbols, which is followed by an "interpretation" of whatever new results may have been generated. De Morgan summarized this attitude in 1849 when he wrote: "No science of symbols can be fully presented to the mind, in such a state as to demand assent or dissent, until its peculiar symbols, their meanings, and the rules of operation, are *all* stated."[44]

De Morgan's insistence that interpretation was the culmination of any complete algebraic investigation guaranteed that algebra would, like geometry, be a descriptive study. From this perspective, legitimate algebra must always be accompanied by a subject matter through which it could be interpreted. The nature of the interaction can be illustrated by a puzzle metaphor which occurred and recurred in De Morgan's work.

A person who puts one of these [a dissected map or picture] together by the backs of the pieces, and therefore is guided only by their forms, and not by their meanings, may be compared to one who makes the transformations of algebra by the defined laws of operation only; while one who looks at the fronts and converts his general

[43] Ibid., p. 174.
[44] Augustus De Morgan, *Trigonometry and Double Algebra*, (London: Printed for Taylor, Walton and Maberly, 1849), p. 89.

knowledge of the countries painted on them into one of a more particular kind by help of the forms of the pieces, more resembles the
investigator and the mathematician.[45]

Thus, when rightly practiced, algebraic manipulations could lead
from general knowledge of a subject matter into a more specific
knowledge of its internal relations. The abstract study of algebra was
valid insofar as it pointed to and aided one to grasp what might otherwise be obscure aspects of the material under scrutiny.

De Morgan's perspective, which allowed for the value of symbolic
manipulation but insisted that ultimately mathematical truth was to
be found in its interpretation, succeeded in saving analytic arguments from being totally discredited within the mid-century English
view of mathematics. It was eloquently argued before the 1852 Parliamentary Commission formed to investigate the Cambridge education. The Commission's investigation of Cambridge mathematics
was a bit tardy since the curriculum had been reformed only four
years earlier. The report is nonetheless useful for showing the issues
surrounding the mathematical curriculum at this time. John Herschel and George Peacock sat on the five-person commission; William Whewell was among those who gave evidence before it. But
these representatives of the Analytical Society did not make up the
whole community involved. There were also younger men who
inherited many of these men's ideas but developed them into a
somewhat different position.

One such representative of the second generation was the young
founder and editor of *The Cambridge Mathematical Journal,* Robert
Ellis. In his evidence before the Commission, Ellis eloquently
argued against the exclusive focus on geometry and for the importance of analysis. He viewed mathematics as a vehicle for discover-
ing and understanding the fundamental principles of a given subject
matter, the "natural relations" among its elements. He argued that
in the search for understanding, unless space were the subject being
considered, there was no essential distinction between analytic and
geometric modes of reasoning.

The principle on which this re-action against the newer methods is

---

[45] Augustus De Morgan, "On the Foundation of Algebra, No. II" (Read Nov. 29, 1841),
Transactions of the Cambridge Philosophical Society, 7 (1842):289–90.

chiefly based, namely that the mind of the Student ought to be as
much as possible conversant with fundamental conceptions is, I
think, perfectly correct. But it does not follow that analytical methods
ought to be discouraged. Demonstrations may be geometrical, and
yet in a high degree artificial; and first principles may be lost sight of
in a maze of triangles, no less than in a maze of equations. Though
in mathematical investigations there is no royal road, yet there is a
natural one, that namely which enables the Student, as far as possible, to grasp the natural relations which exist among the objects of
his contemplation. If this route be followed, it matters but little
whether the reasoning be expressed by one set or kind of symbols or
by another—in plain words—in short hand—or algebraically.[46]

The unique truth of geometry as the study of space was not an
important issue from Ellis's perspective. He maintained that the
function of mathematics was to clarify the principles underlying a
wide variety of fields. He espoused a dynamic, multiperspectival
approach to learning for which the flexibility of analysis was
essential.

It is not by merely fixing in the memory the successive steps of a
single mode of demonstration, or even by studying several, if we
allow them to remain in the mind as distinct and heterogeneous processes of thought, that we are to acquire a complete insight into the
subject in hand, but by a more discursive method,—by inquiring per-
petually into the grounds and reason of what we are doing,—by
interpreting our symbols and following the train of geometrical or
physical conceptions to which their interpretation leads, and again
by retracing our steps and passing from general considerations or
purely geometrical reasoning to the technical language of symbols.
Every change of form should be suggestive of a new aspect of the
subject, and it is thus that the simplest way of considering it is to be
discovered.[47]

Several others who testified before the Cambridge Commission
shared Ellis's view in which both analysis and geometry were important as tools for scientific investigation. Their proper function was

---

[46] Cambridge University Commission, "Evidence from the University," (1852), pp. 223–24.
[47] Ibid., p. 224.

not to sit enthroned as epistemological queen of the sciences but rather to be busy handmaidens who could serve a variety of masters.

Echoes of this opinion can be found in the "Report" of the Cambridge Commissioners. Relying on Ellis's testimony, they concurred that "where relations other than those of space are concerned, what is commonly understood by the geometrical mode of treatment is hardly less symbolic than Algebra itself."[48] Therefore, on the more advanced part of the Tripos, they did not recommend that geometrical proofs be insisted upon. However, in the elementary portion of the examination, which was more directly relevant to the liberal education, the indubitable truth of Euclidean geometry ruled supreme.

The discussions surrounding the 1848 Tripos reform reveal the complexities of the view of mathematics held by Britain's practicing scientists, as well as the conflicts among them. The range of interpretations of mathematics, though large, was not infinite, however; it was consistently bounded by the insistence that mathematics be a truth-seeking enterprise. Thus, even while agreeing that both geometry and analysis might in some situations be used symbolically, the Cambridge Commissioners, like all who testified before them, were careful to emphasize the importance of interpretation for any valid study. They wrote,

> The confessedly superior power of the modern analytical methods makes it indispensable to teach them in their applications to physical subjects, but their tendency undoubtedly is to bridge over the interval between the axiomatic principle and its remote conclusions by processes *sui generis* and which have, or at least appear to have, no relation or (studiously) as little relation as possible to the subject matter. The evil effect of this in giving rise to vagueness of conception in the mind of the Student is obvious—and can only be counteracted by insisting on (what is always possible) their continual interpretation and translation into the language of the subject.[49]

The subjects for mathematical interpretation could be drawn from a wide range of physical investigations, but interpretations were always necessary for a real understanding. Ultimately mathematics

---

[48] "Report of the Cambridge University Commission," p. 113.
[49] Ibid.

had to remain descriptive despite all temptations to move in a more formal direction.

ANALYTICAL DEVELOPMENT AND GEOMETRICAL TRUTH

The conviction that the interpretation of analytic forms could be one route to the attainment of mathematical truth prevented analysis from being totally discredited within the Cambridge education. At more advanced levels of research, however, the descriptive approach to analysis raised its own kind of problems. Geometry was the prime subject matter within which to find interpretations for new analytical results, and Euclid's *Elements* was universally accepted as the definitive treatment of this subject. However, interpreting new analytic results through Euclid was often difficult because many of them seemed to require radically changing classical ideas. The interpretative view of analytical truth was thus often restrictive.

Despite the difficulties they often encountered in trying to geometrically interpret new analytical results, England's mathematicians persisted in trying. Their efforts repeatedly forced them to carefully reconsider and painstakingly delineate the legitimate boundaries of geometrical truth. At times their concerns may appear picayune to the modern reader, but the exercise was crucial for maintaining the special status which had been philosophically and institutionally accorded to mathematics. The central importance of the descriptive view of truth is attested to by the fact that in every significant case where new analytic results pushed the limits of geometrical interpretation, Euclidean space remained the final arbiter of what was valid.

The delineation process can be illustrated by the fate of some ideas suggested in 1839 by the brilliant, though short-lived, Duncan Gregory. In an article entitled "On the Existence of Branches of Curves in Several Planes," Gregory proposed an extension of the concept of the circle. He noted that the analytic form of a circle, $\sqrt{x} = r^2 - y^2$, could have imaginary as well as real values. It was as artificial to ignore those values when interpreting the equation as it would be to exclude negative ones.

Yet, after all, the difference between it $[\sqrt{-1}]$ and the symbol − [minus sign] is not so very great, both admitting of easy interpretation in the science of geometry. . . .

It appears to me, that if we once admit anything beyond what are called positive values of the variables, that is, pure arithmetical values wholly independent of the symbol + [plus sign], there is no reason why we should confine ourselves to − [minus sign]. . . . I therefore hold, that we must either limit ourselves to the one quadrant formed by the positive axes, or we must be prepared to consider the curve as existing in several planes.[50]

On this basis, Gregory proposed extending geometrical conceptions of plane curves to include imaginary values as branches in planes perpendicular to the *xy* plane. William Walton developed these ideas of Gregory's in several additional articles published in 1840 and 1841. In the articles, Walton further explored the three-dimensional forms generated if algebraic expressions in two variables were interpreted generally—that is in ways including imaginary as well as negative values for the variables.[51]

In his initial article Gregory had stressed that for most purposes, the imaginary branches of plane curves were superfluous. "Practically, little attention will be paid to curves existing out of the plane of reference, since the curves themselves do not come sufficiently under our eye to attract much interest," he noted.[52] He recognized, however, that they were centrally important for understanding the relationship between analysis and geometry. The negotiation of this relationship illustrates the impact the English descriptive view of mathematical truth had on mathematical developments.

The unique identifying mark of geometrical truth was that it was at once subjective and objective. To use Herschel's phrase, geometrical truths were embedded in objective space "as the statue in

---

[50] D[uncan] F. Gregory. "On the Existence of Branches of Curves in Several Planes," *The Cambridge Mathematical Journal (CMJ)*, 1 (1839):259.
[51] W. W[alton], On the General Theory of the Loci of Curvilinear Intersection," *CMJ*, 2 (1841):85–91; W. Walton, "On the General Interpretation of Equations between two Variables in Algebraic Geometry," Ibid., pp. 103–113; W. Walton, "On the General Theory of Multiple Points," Ibid., pp. 155–67; W. W[alton], "On the Existence of Possible [Real] Asymptotes to Impossible [Imaginary] Branches of Curves," Ibid., pp. 236–39.
[52] Gregory (1839), pp. 265–66.

the marble." At the same time, they were subjectively solid; once learned, they were known absolutely and with complete conceptual clarity. It was this conceptual clarity which set geometrical truths off from those known in the less perfect sciences. It was the mark of geometry's special truth status, and throughout the century, new geometrical ideas were scrupulously examined under its blinding light. Any suggestions which were not adjudged to be conceptually clear were firmly weeded out and ejected from the pristine sanctuary of legitimate geometry.

This is what happened to Gregory's and Walton's ideas. Their suggestion that there might be extra, "imaginary" curves where previously there had just been simple circles was seen as an intrusion on the territory of geometry where conceptual considerations defined the subject matter. Thus, their investigations soon came under attack in a "Note" appended to George Salmon's *A Treatise on the Higher Plane Curves* published in 1852.

In his "Note," Salmon characterized Gregory's and Walton's approach as follows:

> Strictly speaking, then, there are no plane curves, but every curve, part of whose course may be a plane, is accompanied by branches in space; and the ordinary methods of interpretation, which take no account of these branches in space, give as imperfect a notion of the course of a curve as a system which should take no account of the negative values of the co-ordinates.[53]

To Salmon, this involved changing the delicate balance between the essential subject matter of geometry, which was Euclidean space, and the analytical or deductive forms used to describe it. He elaborated,

> We know what a circle is before we know anything about the equation $x^2 + y^2 = a^2$, and any interpretation of this equation differing either by defect or excess from our previous geometrical conception, must be rejected. We discover that we should be wrong in leaving the sign $-$ uninterpreted, because then the equation $x^2 + y^2 = a^2$ would only

---

[53] George Salmon, *A Treatise on the Higher Plane Curves* (Dublin: Hodges and Smith, 1852), p. 302.

represent a fragment of a circle; and we may in the same manner discover that it is objectionable to give a real interpretation to the symbol $\sqrt{-1}$ in the equations of Analtyic Geometry, because then $x^2 + y^2 = a^2$ represents not only a circle but an irrelevant curve besides.[54]

Further on he emphasized the same point again, writing,

if these curves differ from a circle in form and properties, then it is an abuse of language to speak of them as branches of the circle, merely because they can be represented by the same equation . . . ; it is to confound two distinct ideas, because they can be expressed by the same symbol; it is, in short, no better than a mathematical pun.[55]

Technically Salmon was an Irishman, rather than an Englishman, but his remarks were certainly taken seriously by the Englishmen against whom they were directed. His argument was promptly countered by Walton (Gregory had been dead since 1844) in an article entitled "On the Doctrine of Impossibles in Algebraic Geometry."[56] Augustus De Morgan also joined the fray in an article directly following Walton's which included comments "On the Interpretation of the Equation of a Curve."[57] The discussion appears to have ceased on compromise terms offered in De Morgan's article.

There, De Morgan proposed that a distinction be made between algebraic geometry, where Salmon's spatial structures would define the limits of the investigation, and geometrical algebra, where the equation of the curve would be primary. The choice of which enterprise to pursue was thus left to the individual taste of the investigator. De Morgan clearly revealed his primary interest in geometrical algebra, where geometry was used "in aid of algebra to assist in gaining representations of functions."[58] Nonetheless, he made it

---

[54] Ibid., p. 303.
[55] Ibid., p. 304.
[56] William Walton, "On the Doctrine of Impossibles in Algebraic Geometry," *Cambridge and Dublin Mathematical Journal (CDMJ)*, 7 (1852):234–42.
[57] Augustus De Morgan, "On the Signs + and − in Geometry (continued) and On the Interpretation of the Equations of a Curve," *CDMJ*, 7 (1852):242–51.
[58] Ibid., p. 243.

clear that in algebraic *geometry* notions like that of a circle could not be analytically defined. In geometry, clarity of spatial conception defined the field.

The significance of this discussion is perhaps difficult to see because of its relatively esoteric nature. Salmon's insistence that circles be defined by our conception of them seems reasonable from a pragmatic standpoint if nothing else. De Morgan's compromise seems adequate, but almost trivial. The importance of the issues lurking in the argument can perhaps be seen more clearly in the larger discussions of spaces with more than three dimensions— higher dimensional spaces—which drew attention throughout the rest of the century. In this context, the geometrical importance of the subjective criteria of clear conception, as opposed to formal mathematical considerations, can be seen clearly.

During the second half of the century, mathematicians were turning with increasing frequency to analytically suggested higher dimensional space in order to solve problems. In an article entitled "On Some Points in the Theory of Elimination" published in 1866 in the *Quarterly Journal of Pure and Applied Mathematics,* Salmon suggests the terms on which these arguments were used. Here Salmon wrote,

> I have already completely discussed this problem when we are given three equations in three variables. . . . The question now before us may be stated as the corresponding problem in space of p dimensions. But we consider it as a purely algebraical question, apart from any geometrical considerations.
>
> We shall however retain a little of geometrical *language,* both because we can thus avoid circumlocutions, and also because we can thus more readily see how to apply to a system of p equations, processes analogous to those which we have employed in a system of three.[59]

Another example of the use of higher dimensional spaces can be found in Arthur Cayley's 1869 "Memoir on Abstract Geometry." Here he wrote,

---

[59] George Salmon, "On Some Points in the Theory of Elimination," *The Quarterly Journal of Pure and Applied Mathematics,* 7 (1866):327–28.

The science [of Abstract Geometry] presents itself in two ways;—as a legitimate extension of the ordinary two- and three-dimensinal geometries; and as a need in these geometries and in analysis generally. In fact whenever we are concerned with quantities connected together in any manner, and which are, or are considered as variable or determinable, then the nature of the relation between the quantities is frequently rendered more intelligible by regarding them (if only two or three in number) as the co-ordinates of a point in a plane or in space: for more than three quantities there is, from the greater complexity of the case, the greater need of such a representation; but this can only be obtained by means of the notion of a space of the proper dimensionality; and to use such representation we require the geometry of such space.[60]

In these cases, both Salmon and Cayley tried to justify their use of geometrical argument to consider problems which did not have straightforward, three-dimensional interpretations. Salmon was clearer than Cayley that by considering problems in p dimensions he was just using a figure of speech; he was careful not to make any claims for the reality, whether conceptual or physical, of such spaces. Cayley was less explicit on this point and went so far as to refer to the "notion of a space of [higher] dimensionality." Nonetheless, it is clear from the rest of his work that Cayley would not have defended a view which would have put his higher dimensional spaces on the same epistemological level as the readily conceivable Euclidean space of three dimensions.

Neither Salmon nor Cayley would have considered the possibility that spaces of higher dimension might be real in the same sense that Euclid's was, and there were few who would seriously have disagreed with them. The issue was specifically argued in the wake of a presidential speech to the mathematics and physics section of the BAAS by the somewhat more daring J. J. Sylvester. In 1869, Sylvester suggested that instead of considering "the alleged notion of generalised space as only a disguised form of algebraical formulisa-

---

[60] Arthur Cayley, "A Memoir on Abstract Geometry," (Read Dec. 16, 1869), *Philosophical Transactions of the Royal Society of London*, 65 (1870):51–63; reprinted in Arthur Cayley, *Collected Mathematical Papers*, vol. 6 (Cambridge: University Press, 1893), p. 456.

tion," it be considered as a legitimate part of geometry.[61] His claim for their admission to this status was not mathematical, however. He did not follow Gregory in the suggestion that algebraic forms be used to define geometrical truth, for example. Instead he used the criteria of conceivability, which Whewell had suggested was the hall-mark of geometrical truth, as the way to discriminate those things which were legitimately part of the concept of space, and hence of geometry, and those which were not.

Sylvester's route was rather circuitous, however. Instead of claim-ing that *he* could conceive of them, he pointed to a number of his colleagues, including Cayley, Salmon and William K. Clifford, who, he asserted, had this capacity. Backed by the alleged powers of these luminaries he suggested that the "transcendental space" of more than three dimensions be accepted as real and a legitimate part of geometry.[62]

The complete subjectivity of the conceivability criteria on which Sylvester judged what was and what was not part of geometry can be seen in the response to Sylvester's claim. Within two weeks, his allegations were challenged by the Shakespearian scholar, C. M. Ingleby. Ingleby relied on Kant to support his assertion that an essential part of the concept of space was that it had only three dimensions. He countered Sylvester's conceivability claim with a statement from his friend Salmon, whom he quoted as saying: "I do not profess to be able to conceive of *affairs* of four dimensions. . . . I advise you to believe whatever Sylvester tells you, for he has the power of seeing things invisible to ordinary mortals."[63] Since neither Salmon nor Sylvester, nor presumably the other men on whose pow-ers Sylvester had rested his claim, would actually claim to be able to conceive of higher dimensional spaces himself, Ingleby was able to rest comfortably in his conviction that space had exactly three dimensions.

Ingleby was supported in his belief that higher dimensional spaces were not a legitimate part of geometry by virtually all of the English

---

[61] J. J. Sylvester, "Address to the Mathematics and Physics Section," *Report of the Thirty-ninth Meeting of the BAAS held at Exeter in August, 1869,* (London: John Murray, 1870), pp. 1–9; reprinted as "A Plea for the Mathematician," *Nature,* 1 (1869–70):238.
[62] Ibid.
[63] C. M. Ingleby, "Transcendent Space," *Nature,* 1 (1869–70):289.

mathematicians who commented on this subject during the ensuing decades. Their focus on epistemological or psychological criteria to judge mathematical validity was universal, although they sometimes differed in their interpretations of the human psyche. Thus, the German physiologist, Hermann von Helmholtz, who was so central to English discussions of geometry at this time as to arguably be English in this respect, was an empirical psychologist. He emphasized that the perceptual apparatus was critical in transmitting our knowledge of space. Accordingly he stated: "As all our means of sense-perception extend only to space of three dimensions, and a fourth is not merely a modification of what we have but something perfectly new, we find ourselves by reason of our bodily organization quite unable to represent a fourth dimension."[64] William Clifford, who emphasized spatial experience rather than perception as the source of spatial conception, concluded from the impossibility of moving three-dimensional objects into the fourth dimension: "We arrive, then, at the result that *space is of three dimensions*."[65] In short, Samuel Roberts, president of the London Mathematical Society, seems to have been accurate when he asserted in 1882, "It is admitted, on all hands, that we can form no conception whatever of a fourth geometrical dimension."[66]

The whole issue of the conceivability of higher dimensional spaces took center stage in the consideration of their geometrical legitimacy. It totally eclipsed the kind of algebraic perspective suggested by Gregory, in which algebraic forms would define geometrical concepts. It also overshadowed another perspective on the whole issue

---

[64] H[ermann von] Helmholtz, "The Origin and Meaning of Geometrical Axioms," *Mind*, 1 (1876):318–19.

[65] W[illiam] K[ingdon] Clifford, "The Philosophy of the Pure Sciences, Pt. II: The Postulates of the Science of Space," *Contemporary Review*, 25 (1875):365.

[66] Samuel Roberts, "Remarks on Mathematical Terminology, and the Philosophic Bearing of Recent Mathematical Speculations concerning the Realities of Space." (Read Nov. 9, 1882), *Proceedings of the London Mathematical Society*, 14 (1882–3):12. A highly original and rare movement away from this position in which geometrical conception legislates against higher dimensionality is in Edwin Abbott Abbott's *Flatland: A Romance of Many Dimensions*, (London: Seeley, 1884). In this fictional story about dimensionality, Abbott undercut the strict separation of three-dimensional from higher dimensional spaces by casting doubt on the clarity of the three-dimensional conception. Whereas for Whewell three-dimensional conceptions were clear and indubitable, in *Flatland* the main characters struggle continuously to maintain even the shadows of these conceptions.

suggested in a paper published by Julius Plücker in 1865 in the *Philosophical Transactions of the Royal Society.*[67] In this work, the German had demonstrated that regular three-dimensional space could be interpreted as four-dimensional, if lines rather than points were taken as the fundamental unity. Thus, he provided a three-dimensional model for four-dimensional (and in fact *n*-dimensional) space. From a modern perspective this paper is considered as a pivotal work in the development of higher dimensional geometries.

In nineteenth-century England, however, Plücker's demonstration was known but not considered relevant to the crucial question about the reality of higher dimensional geometries. One of the rare instances in which Plücker's work was even considered in this context is William Spottiswoode's 1878 Presidential Address to the British Association for the Advancement of Science. Here Spottiswoode first introduced higher dimensional spaces as mere analytical artifacts. He then pointed out, "There is, however, another aspect under which even ordinary space presents to us a four-fold, or indeed a mani-fold character," and developed Plücker's interpretation in detail.[68] This discussion did not serve to establish the similarities between these higher dimensional spaces and Euclidean space, however. Spottiswoode concluded his discussion as follows:

> This is in fact the whole story and mystery of manifold space. It is not seriously regarded as a reality in the same sense as ordinary space; it is a mode of representation, or a method which, having served its purpose, vanishes from the scene. Like a rainbow, if we try to grasp it, it eludes our very touch; but, like a rainbow, it arises out of real conditions of known and tangible quantities, and if rightly apprehended it is a true and valuable expression of natural laws, and serves a definite purpose in the science of which it forms a part.[69]

Thus, for Spottiswoode, Plücker's work showed that the language used to talk about space might change considerably, and many dif-

---

[67] Julius Plücker, "On a New Geometry of Space," *Philosophical Transactions of the Royal Society,* 155 (1865):725–91.

[68] William Spottiswoode, "Presidential Address, " *Report of the Forty-eighth Meeting of the BAAS held at Dublin in August 1878.* (London: John Murray, 1879), p. 21.

[69] Ibid., pp. 22–23.

ferent angles might be explored in considering it. From this point of view, higher dimensional language might be interesting and helpful in deepening spatial understanding. Ultimately, however, three-dimensional space was special because it was a reality—more or less clearly grasped conceptually—which existed independently of geometry, the science which described it. Throughout the mathematical discussions of higher dimensional space, this conceptual reality—so important to the descriptive interpretation of mathematics defended within the British educational and philosophical traditions—remained firm.

William Kingdon Clifford. *(Courtesy of the Royal Society, London.)*

James Joseph Sylvester. *(Courtesy of the Ferdinand Hamburger, Jr Archives of the Johns Hopkins University.)*

CHAPTER 2

Non-Euclidean Geometry and Mathematical Truth

Until late in the seventh decade of the century, all English formulations of the nature of geometry were constructed from Euclidean geometry. The space described by Euclid's *Elements* defined the subject. Interpreting the kind of understanding revealed in this classic work was the challenge which faced the epistemologists; passing it on to future generations was the task of the educators; defending it against illegitimate encroachments was the job of the careful mathematicians. Although there was some criticism of its logical detail,[1] the geometry presented in Euclid's *Elements* was essentially unassailed in the first part of the century. The integration of geometry into the larger cultural scheme of things was based upon this immovable foundation.

Even as the implications of this fixed point in the intellectual landscape were being mapped out, however, undermining forces were at work. Unbeknownst to the English community, at the beginning of the century several continental mathematicians had carefully scrutinized Euclid's work and decided that it was not adequate. They asserted that its insights did not include all spatial knowledge. In

---

[1] Augustus De Morgan wrote a detailed critique of the logical structure of Euclid's work in "Short Supplementary on the First Six Books of Euclid's Elements," *The Companion to the Almanac*, (London: Charles Knight, 1849), pp. 6–20.

61

fact, they suggested that other spaces were also possible and began to explore what are now generally called non-Euclidean geometries. Taken by itself, the term non-Euclidean geometry would appear to encompass any geometrical development which is not included in Euclid. Usually, though, it is confined to a specific set of investigations which focused on alternatives to the Euclidean theory of parallels. These are the developments which were eroding the ground under Euclid's geometry even as the English were painstakingly constructing an epistemological world-view upon it.

The nineteenth-century study of non-Euclidean geometry represented the first significant departure from the Euclidean view of geometrical space in almost two and a half millennia. At the same time, non-Euclidean geometries can be seen as the culmination of a long tradition which questioned Euclid's interpretation of parallel lines. Almost from the time of the first appearance of Euclid's *Elements*, the fifth of the postulates on which he based his system was the object of intermittent concern. Through the centuries there were numerous attempts to improve on Euclid's treatment.

The underlying problem was that on the subject of parallels, Euclid was strikingly inelegant. Whereas the other postulates, such as the fourth which maintained "That all right angles are equal to one another," were utterly simple and therefore could easily be defended as self-evident, the fifth was different. Euclid enunciated this troublesome postulate as follows:

That, if a straight line falling on two straight lines make the interior angles on the same side less than two right angles, the two straight lines, if produced indefinitely, meet on that side on which are the angles less than the two right angles.

When combined with the definition of parallel lines—"Parallel straight lines are straight lines which, being in the same plane and being produced indefinitely in both directions, do not meet one another in either direction"[2]—this postulate formed the basis for Euclid's theory of parallels.

---

[2] *The Thirteen Books of Euclid's Elements,* Introduction and Commentary by Thomas L. Heath, 2nd ed. rev., 3 vols. (New York: Dover Publications, 1956), 1, pp. 154–55.

However, the Greek was not compelled to use the fifth postulate until the proof of his twenty-ninth proposition. Furthermore, the twenty-ninth proposition was simply the converse of the twenty-eighth, which was proved without the awkward postulate. That so much could be proved without it, and that, when finally called upon, its use was relatively minor, further contributed to the perception that the fifth postulate was unsatisfactory and probably unnecessary. For centuries men tried to derive it from the four simpler, more clearly self-evident postulates which preceded it. However, their attempts to construct a direct proof always failed.

In the beginning of the eighteenth century an Italian Jesuit, Girolamo Saccheri, took a novel approach to the centuries-old problem of proving the parallel postulate which opened up a vast new territory of mathematical innovation. In 1733, Saccheri published a work entitled *Euclides ab omni naevo vindicatus,* in which he approached the problem of proving Euclid's fifth postulate by an indirect route. Rather than attempting to prove the postulate true, he assumed that it was *not* true and tried to demonstrate the falsity of any alternative postulate.

To effect his proof, Saccheri relied on a plane quadrilateral, *ABCD*, in which the opposite sides *AC* and *BD* were assumed to be equal and perpendicular to a third side *AB* (figure 1).

It was easy to prove that angle *C* and angle *D* were equal to one another, but that they were right angles posed a more difficult problem. Here Saccheri identified three possibilities: that the angles be

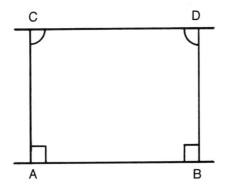

Figure 1. Saccheri's Quadrilateral

both right, both obtuse (greater than right), or both acute (less than right). He then proceeded to develop the consequences of the assumptions of the acute and obtuse angles, confident that they would generate contradictions.

Using the assumption of the obtuse angle, Saccheri quickly found the contradiction he sought. His proof is too technical to enter in detail, but the gist of the argument can be illustrated rather easily. If angles *C* and *D* were both larger than right angles, the line *CD* would act as if it were slightly bowed downward (figure 2). Eventually it would intersect *AB* not only in one but in two places. Clearly, without a radical adjustment in the definition of parallels, such a line could not be called parallel to *AB*. Saccheri was therefore able to firmly rule out this hypothesis.

In the case of the acute angle, however, generating a contradiction was more difficult. As figure 3 illustrates, viewing line *CD* as slightly bowed upward would make angles *C* and *D* both less than right angles. With the line bowed this way, there is no obvious problem with calling *CD* parallel to *AB* since the two lines show no sign of intersecting each other. Furthermore, as Saccheri discovered after much effort, it is hard to find any other straightforward reason to reject this hypothesis. As he tested it in various ways, Saccheri repeatedly failed to generate a clear contradiction. Finally he

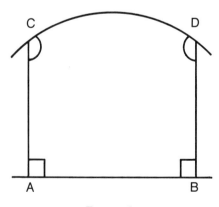

FIGURE 2.

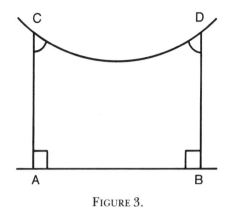

FIGURE 3.

resorted to rejecting it categorically, on essentially intuitive grounds: "The Hypothesis of the Acute Angle is absolutely false; because repugnant to the nature of the straight line. (Proposition XXXIII)"[3]

Despite his conviction that his work vindicated Euclid, Saccheri generated a number of results which were later incorporated into nineteenth-century non-Euclidean geometry. For example, his work contains the result that if the sum of the angles of a triangle is less than two right angles in one triangle, it must be so in all. In fact, except for the final proposition, in which he denounced what he had developed before as essentially flawed, much of Saccheri's work is indistinguishable from the work of non-Euclidean geometers.

Almost a century elapsed between 1733, when Saccheri published his work, and the first publications of what are now generally recognized as non-Euclidean geometries. In the interim numerous mathematical investigators probed the foundations of geometry, focusing on the problems raised by the fifth postulate. Some, like Adrien Marie Legendre, approached the problem directly and tried to prove the postulate as a theorem. Others, like Johann Heinrich

---

[3] *Girolamo Saccheri's Euclides Vindicatus,* edited and translated by George Bruce Halsted, (Chicago: Open Court, 1920), p. 173.

Lambert, followed Saccheri in approaching the problem indirectly and searched for inconsistencies in the hypotheses of the acute and obtuse angles.[4] Many parts of non-Euclidean geometry were developed and clarified in these works. Legendre presented much of what was masked in Saccheri in a clear and readable fashion; Lambert was apparently the first to notice that under the hypothesis of the acute angle, the defect of the angle sum of a triangle (the amount that its angle sum differed from two rights) was proportional to the area. Similarly, Lambert pointed out that the same hypothesis led to an absolute standard for length.

However, neither of these eighteenth-century thinkers seriously entertained the possibility that the theorems they were generating from the acute angle hypothesis might describe space as well as Euclid's geometry did. Legendre was mistakenly satisfied with a proof of Euclid's postulate as a theorem. Lambert's work, only published posthumously in 1786, was ultimately inconclusive. Although his failure to generate an inconsistency suggested that Euclid's system might not constitute the only possible geometry, Lambert did not commit himself to this position. Instead he self-consciously left the question open for further consideration.

In the opening decades of the nineteenth century, several men went further than these eighteenth-century investigators by considering that the non-Euclidean geometry generated from the acute angle hypothesis might be as true as Euclid's. They wondered whether Euclid's geometry really constituted the absolutely true description of space. This suggestion contained a challenge to the interpretations of spatial knowledge which had been attendant on geometrical investigation throughout the eighteenth century. It is this challenge, as opposed to the specific theorems they developed, which distinguished these "non-Euclidean geometers" from their eighteenth-century predecessors.

The beginning of non-Euclidean geometry is traditionally traced to the work of two otherwise obscure mathematicians, János Bólyai

---

[4] J[eremy] J. Gray and Laura Tilling, "Johann Heinrich Lambert, mathematician and scientist, 1728–1777," *Historia Mathematica*, 5 (1978):13–41.

and Nicholai Lobachevskii. In the 1820s, working independently, each of these men essentially took Saccheri's indirect approach to proving the fifth postulate. They assumed different postulates in place of Euclid's parallel one and derived geometrical systems from these alternative constructions. Unlike many of their eighteenth-century predecessors, however, these two did not conclude that they were on the wrong track as they developed the results of their non-Euclidean postulates. Nor did they ignore the question of whether Euclid's postulate could be proved from the others. They concluded that the fifth postulate could not be proved from the rest of Euclid's system, and therefore that contradictory assumptions could be advanced equally cogently.

From a certain perspective, one can argue that Bólyai's and Lobachevskii's conclusions vindicated Euclid, because the Greek had maintained the fifth postulate without proof, rather than attempting to place it among his theorems. It seems that the Greek recognized that this postulate was independent of the others and therefore not susceptible to proof by them. This interpretation does not do justice to the vision with which the early non-Euclidean geometers approached their work, however. The difficulty which led them to investigate the status of the fifth postulate lay in its recalcitrant clumsiness, which made it difficult to accept it as self-evident or even compellingly persuasive. The impetus behind the repeated investigations of the parallel postulate was a nagging doubt about the truth of a geometry constructed on such a conceptually flimsy postulate.

In an important sense, the very existence of the doubt was testimony to the inadequacy of the postulate as a cornerstone for the geometrical study of space. When Lobachevskii and Bólyai concluded that this non-evident postulate could not, in fact, be derived from the others, they were immediately led to question whether the geometry Euclid had developed did, as their contemporaries supposed it did, truly describe the only true or possible spatial reality. A critical factor suggesting this problem was the fact that the indirect investigation of the parallel postulate entailed developing an equally consistent alternative geometry in which Euclid's theory of parallels was not valid. Neither Lobachevskii nor Bólyai was content to conclude by merely asserting that Euclid was right to leave the postulate unproved. Both immediately raised the question of which geometry, Euclid's or the newly elaborated alternative, served to describe the space of common experience. Bólyai commented toward the end of his paper, "thus far it has remained undecided, which of these two [geometries—his or Euclid's] has place in reality,"[5] and Lobachevskii concluded his work with the observation:

> The equations [of non-Euclidean trigonometry] attain for themselves already a sufficient foundation for considering the assumption of imaginary geometry [geometry under the assumption of the acute angle] as possible. Hence there is no means, other than astronomical observations, to use for judging the exactitude which pertains to the calculations of the ordinary geometry.[6]

Lobachevskii and Bólyai were excellent mathematicians, using the term in a relatively narrow sense. Their work was developed in the context of long-standing mathematical tradition.[7] At the same time, however, to understand their historical position it is necessary to recognize that both Lobachevskii and Bólyai tried to understand the meaning of their results in broader terms. Their interests were not held within the narrow confines of technical mathematics *per se* but spilled out into the wider intellectual arena.

The significance of their work was considerable, and they both recognized it, as did a small number of their more astute contemporaries. The great geometer Carl Friedrich Gauss apparently suppressed his similar researches during this period to avoid stirring up controversy and the "chatter of the Boeotians." In the event, however, it seems that no one had to fear Boeotian chatter. One of the

---

[5] J[ohannes Bólyai], John Bolyai, *The Science of Absolute Space,* translated by George Bruce Halsted, Supplement to Roberto Bonola, *Non-Euclidean Geometry: A Critical and Historical Study of its Development,* authorized translation with appendices by H. S. Carslow, (New York: Dover, 1955), p. 47.

[6] [Nicholai Lobachevskii], Nicholas Lobachevsky, *Geometrical Researches on the Theory of Parallels,* translated by George Bruce Halsted, Supplement to Bonola, *Non-Euclidean Geometry,* pp. 44–45.

[7] Jeremy Gray, "Non-Euclidean Geometry—A Re-Interpretation," *Historia Mathematica,* 6 (1979):236–58.

most striking things about Lobachevskii's and Bólyai's work is the deadening silence with which it was received in the first half of the century.

In England, the silence can be attributed largely to ignorance of the work of these obscure Eastern European mathematicians. The ignorance might in turn be explained as a lack of interest in or understanding of the subject. Before late in the 1860s, there is little indication that any English mathematicians seriously considered the possibility that Euclid's fifth postulate could be doubted. Aside from isolated attempts to make the postulate more persuasive and self-evident,[8] and the inevitable efforts to prove it from the other four more self-evident axioms,[9] there was little mathematical discussion of the parallel postulate in England.

There was, however, some philosophical discussion which in retrospect has appeared to be relevant to non-Euclidean issues. In his eighteenth-century analysis of the relations between perceptual sign and the thing signified, Bishop Berkeley had pointed out that the space which one visually perceived was not that described by classical geometry. The source of visual spatial perception is the retina, which is flat and two-dimensional. A significant interpretative process is involved in turning the retinal image into the infinite three-dimensional space which humans see.

Berkeley's observation that the space imaged on the retina was not the same as the space which was visually perceived led directly to the epistemological tradition presented in the last chapter. It pointed irresistibly to the problems inherent in judging how we know the world around us. Euclidean geometry provided the basic starting point for these discussions. It was the spatial concept which provided the interpretative link between incomplete retinal experi-

---

[8] Notable among these attempts was T. Perronet Thompson's *Geometry without Axioms, or the First Book of Euclid's Elements*, 4th ed., (London, 1833), which was reviewed in 1834 by Augustus De Morgan ("Review of Geometry without Axioms . . . Fourth edition," *Quarterly Review of Education*, 7 (1834):105–15). Thompson derived the parallel postulate from relations of spherical geometry, an attempt which De Morgan felt did not render it more self-evident. In Duncan M. Y. Sommerville, *Bibliography of non-Euclidean Geometry including the Theory of Parallels, the Foundations of Geometry and Space of n Dimensions*, (London: Harrison & Sons, 1911), there are further references to Thompson's later attempts along these lines.

[9] See Sommerville, *Bibliography*, for references to works of this kind.

ence and complete spatial perception. Thus, this concept played an absolutely vital role in human natural knowledge. Its nature and origins were avidly discussed in an attempt to gain a better understanding of how humans learn what is true.

Recently it has been noted that there was another, smaller tradition which grew out of Berkeley's insight into the disparity between visual experience and perception. Thomas Reid, a Scottish philosopher, took some time from his consideration of the commonly perceived spatial concept to focus on the geometry of the retinal image. He tried to construct a geometry of the retina which would describe the spatial impressions of a single, immobile eye. He developed a series of axioms for this geometry, which he explicitly recognized were different from Euclid's. Furthermore, he elaborated a complicated model of the perceptual world which these axioms described, carefully explaining the experiences of an imaginary group of "Idomenians" whose only spatial experience would be visual.[10]

Reid's geometry was structurally non-Euclidean. His presentation of it did not challenge the truth of the Euclidean geometry, however, which describes the space in which human beings live and move. It served more as an example of the way a distorted experience could lead to an inadequate understanding of Euclidean truth than as a challenge to that truth. Thus Reid's work stands as a fascinating contribution to the tradition of eighteenth-century forerunners to non-Euclidean geometry, although it lacked the philsophical impact of the nineteenth-century investigations.

Nonetheless, Reid's work defined a small but significant tradition in England. His suggestion that an alternative geometry of the retina could be constructed was intriguing enough that there were occasional other attempts to describe the geometries generated by distorted experiences.[11] It was also disturbing enough to Whewell's

---

[10] Norman Daniels has focused on Reid's mathematical construction as a hitherto unrecognized non-Euclidean geometry. Norman Daniels, *Thomas Reid's Inquiry: The Geometry of Visibles and the Case for Realism,* with a Foreword by Hilary Putnam, (New York: Burt Franklin and Co., 1974).

[11] See, for example, the article directed against Mansel in [James Fitzjames Stephen], *Essays by a Barrister,* (London: Smith, Elder and Co., 1862).

notions of the necessity of geometrical truth to warrant a rebuttal in the *Philosophy of the Inductive Sciences.*[12]

In some sense, the apex of this tradition of speculating about alternative spatial experiences is a long and detailed paper John Kelland delivered in 1863 to the Royal Society of Edinburgh. Kelland's paper, entitled "On the Limits of our Knowledge respecting the Theory of Parallels," apparently grew out of several years of seminar discussions investigating Reidian kinds of alternative geometries. It also contains one of the earliest English references to the European non-Euclidean tradition. In it Kelland noted, "Some years ago, there appeared in *Crelle's* Journal, a notice of a work entitled 'Imaginary or Impossible Geometry,' viz., a discussion of the conclusions which would follow from the assumption as an axiom, of the hypothesis that 'the three angles of a triangle are together less than two right angles.'"[13] He continued with the admission that he had never seen this work (clearly Lobachevskii's book), but had attempted to develop what results he could from Lobachevskii's assumption. His development, which seems to have been written with virtually no knowledge of the work of Lobachevskii, Gauss or Bólyai, fills the rest of the paper.

Not surprisingly, Kelland failed to demonstrate the impossibility of the acute angle hypothesis in his paper. His conclusion from this failure was not a radical rejection of the truth of Euclid's system, however. Rather, as his title suggests, Kelland was concerned with the *limits* of our understanding of the nature of parallels. He used his development of the acute angle hypothesis as an indicator of places where further work should be focused. Thus, for example, he pointed out that a postulate affirming that "a triangle of which the vertex is given may be increased indefinitely by increasing its base" would be adequate to prove that "the three angles of a triangle

---

[12] William Whewell, *Philosophy of the Inductive Sciences*, 2nd ed. rev., 2 vols, (London, 1847; reprint ed., London: Frank Cass & Co., Ltd., 1967), 1, pp. 101–10. Whewell argued that the single eye would have to rotate in order to collect its impressions, and the knowledge gleaned from properties of this motion would be sufficient to generate an accurate spatial understanding.

[13] Professor [John] Kelland, "On the Limits of our Knowledge respecting the Theory of Parallels" (Read Dec. 21, 1863), *Transactions of the Royal Society of Edinburgh*, 23 (1861–64):434.

taken together [are] equal to two right angles." In this case he noted that "the postulate . . . is not legitimate as a logical canon." It was nonetheless useful to consider such postulates because they suggested novel approaches to the theory of parallels. For Kelland, equally significant was this postulate's tiny scope: "it evidences . . . the extreme narrowness of the limits towards the received doctrine to which the inquiry is pushed."[14] This specificity reassured Kelland that the problem was almost solved, its limits very narrow indeed.

Kelland's treatment is essentially in the spirit of Reid's Idomeniac fantasy or of the investigations of eighteenth-century mathematicians. Like these forerunners he drew no radical conclusions about the truth of Euclid's geometry from his development. He treated his failure to disprove the alternative hypothesis as simply that, a failure. He did not regard it as a definitive indication of any kind about the truth of Euclid's fifth postulate.

A rather different treatment of non-Euclidean geometry can be found in another paper which appeared two years later. Arthur Cayley published a paper, entitled "Note on Lobatchewsky's Imaginary Geometry," in the *Philosophical Magazine* in 1865. Cayley's work is more mathematical and less speculative than Kelland's. However, he revealed only a slightly greater familiarity with Lobachevskii's ideas than the Scot had. He referred to the "curious paper, 'Géométrie Imaginaire' by N. Lobatchewsky, Rector of the University of Kasan, Crelle, vol. xvii, (1837) p. 295–320."[15] Cayley briefly presented a form of the equations Lobachevskii had derived for the trigonometry of his imaginary geometry, and pointed out that if in a triangle the sum of the angles is considered larger than 180°, one had merely the equations of spherical trigonometry. As for Lobachevskii's explanation of the case when the sum of these angles is *less* than 180° (the hypothesis of the acute angle), Cayley wrote:

The view taken of them by the author is hard to be understood. He mentions that in a paper published five years previously in a scientific

---

[14] Ibid., p. 444.
[15] Arthur Cayley, "Note on Lobatchewsky's Imaginary Geometry," *Philosophical Magazine*, 29 (1865):231–33. Reprinted in *The Collected Mathematical Papers of Arthur Cayley*, 11 vols., (Cambridge: University Press, 1889–97), 5:472.

journal at Kasan, after developing a new theory of parallels, he had endeavored to prove that it is only experience which obliges us to assume that in a rectilinear triangle the sum of the angles is equal to two right angles, and that a geometry may exist, if not in nature at least in analysis, on the hypothesis that the sum of the angles is less than two right angles . . . I do not understand this; but it would be very interesting to find a *real* geometrical interpretation of the last-mentioned system of equations.[16]

Despite Cayley's obvious interest in Lobachevskii's analytical development of the formulae of the trigonometry of a space of negative curvature, it had no immediate implications for his view of space. This is not surprising in view of the relationship between analysis and geometry worked out in the discussion of Gregory's and Walton's interpretations of equations of plane curves. In terms of the compromise sketched out by De Morgan, mere analysis carried no necessary implications for the study of space. In his 1865 paper, Cayley was true to the spirit of that compromise. He was intrigued by the challenge of finding an interpretation for the strange equations within the Euclidean spatial concept; "it would be very interesting to find a *real* geometrical interpretation of the . . . system of equations." Clearly, however, he saw no connection between the analytical development and a need to radically reconsider the nature of classically conceived space. Though entertaining, Lobachevskii's ideas merited no more than a "Note."

The peaceful ignorance of European non-Euclidean ideas, reflected in Kelland's and Cayley's early articles, was soon shattered. The 1860s witnessed an explosion of continental interest in non-Euclidean ideas. Gauss's correspondence with Schumacher on the subject was published between 1860 and 1863.[17] In 1866, J. Houël published a French translation of Lobachevskii's book.[18] In addition, new works appeared. Bernhard Riemann's "Habilitationsvortrag," delivered to the faculty of the University in Göttingen, June

---

[16] Ibid.

[17] C. F. Gauss and H. C. Schumacher, *Briefwechsel zwischen K. F. Gauss und H. C. Schumacher*, (Altona: C. A. F. Peters, 1860–65).

[18] N. I. Lobachevskii, *Études géometriques sur la théorie des parallèles, suivi d'un extrait de la correspondence de Gauss et de Schumacher*, translated by J. Houël, (Paris: Gauthier-Villars, 1866).

10, 1854, was first published in 1866.[19] The non-Euclidean ideas in this paper were promptly digested and re-presented in a more comprehensible form by the German physiologist Hermann von Helmholtz.[20] The subject which had lain dormant for the first part of the century came to life in its seventh decade.

This non-Euclidean explosion was quickly felt in England. In the same 1869 address in which he had speculated about higher dimensional spaces, Sylvester referred to Riemann's paper published just three years before. In 1870, Helmholtz began publishing his non-Euclidean thoughts directly in *The Academy*.[21] William Clifford translated Riemann's paper and, in 1873, published it in *Nature*.[22] In 1874, he gave a popular digest of Riemann's ideas in a series of lectures at the Royal Institution.[23] By the mid-1870s non-Euclidean ideas were readily available in England, and the halcyon assurance of Kelland's and Cayley's early ruminations had been broken.

This chapter will focus on the immediate English response to the

---

[19] Bernhard Riemann, "Über die Hypothesen, welche der Geometrie zu Grunde liegen," *Abhandlungen der Koniglichen Gesellschaft der Wissenschafter zu Göttingen*, 13 (1867):132–52.

[20] Hermann von Helmholtz, "Über die thatsälichen Grundlagen der Geometrie," *Verhandlungen des naturhistorisch-medicinischen Vereins zu Heidelberg*, 4 (1866):197–202; "Appendix," *Verhandlungen*, 5 (1869):31–32; Hermann von Helmholtz, "Über die Thatsachen die der Geometrie zum Grunde liegen," *Nachrichten von der Königlichen Gesellschaft der Wissenschaften zu Göttingen*, (1868):193–222.

The date of the first of these articles is difficult to establish. Koenigsberger's biography of Helmholtz contains a letter from Helmholtz to Schering dated 12 April 1868, in which Helmholtz makes it clear that he has just then heard of Riemann's work. Since in his "1866" paper Helmholtz specifically mentions Riemann, there seems to be a discrepancy. Going to the paper itself in the fourth volume of *Verhandlungen des naturhistorisch-medicinischen Vereins* does not definitively fix the date of publication. This volume covers the period from March 1865 to October 1868. Helmholtz's paper is printed in a section surrounded by papers delivered in 1868. Helmholtz's paper is dated 22 May 1866, but the ones immediately preceding and following it are both dated 22 May 1868. This suggests that what is at issue is a misprint, and that the 22 May 1866 date given in the *Wissenschaftliche Abhandlungen* is a perpetuation of this misprint.

However, the misprint theory seems to be confuted by the appearance of a French translation of this paper in the *Société des Sciences physiques et naturelles de Bordeaux*. This translation by Houël is dated 1867, which would suggest the earlier date of 1866 for Helmholtz's article.

[21] Hermann von Helmholtz, "The Axioms of Geometry," *The Academy*, 1 (1870):128–31.

[22] Bernhard Riemann, "On the Hypotheses which Lie at the Bases of Geometry," *Nature*, 8 (1873):14–17, 36–37. Unless otherwise noted, translations are from the Clifford version.

[23] W[illiam] K. Clifford, "The Philosophy of the Pure Sciences, Pt. 2. The Postulates of the Science of Space," *Contemporary Review*, 25 (1875):360–76.

non-Euclidean ideas which were being introduced from the continent. The first section will elaborate some of the mathematical developments at the base of the new wave in geometry. In addition to Lobachevskii and Bólyai, there were two later contributors to non-Euclidean thought: the intense mathematical thinker, Riemann, who died in 1866, but left a rich legacy in his densely packed "Habilitationsvortrag," and the brilliant, charismatic Helmholtz, who cut a path through the thicket of Riemann's work and interpreted it for Europe's intellectuals. These Germans' approach to non-Euclidean geometry was significantly different from the one taken by Lobachevskii and Bólyai earlier in the century. The difference has long been recognized by applying the term "synthetic" to Lobachevskii's and Bólyai's work and "metric" to that of Riemann and Helmholtz.

Early in the 1870s, both "synthetic" and "metric" non-Euclidean ideas were being widely discussed in England for the first time. They raised major issues within the framework in which geometrical and analytical study had been embedded. Most central were questions about the nature of truth: mathematical truth, scientific truth and humanly accessible truth in general. The issues were posed in several ways which reflected significantly on interpretations of the nature of geometrical conception. The second section will focus on three discussions of the nature of mathematical truth which grew out of non-Euclidean ideas.

The third and final section will consider these mathematical discussions in the larger context of English intellectual culture. Mathematical study was as critical a part of the later Victorian construction of knowledge as it had been in the time of the 1848 Tripos reform. However by the 1870s the intellectual situation was different. The community was gripped in a major crisis whose roots are often traced to the publication, in 1859, of Darwin's *On the Origin of Species*. The discussions surrounding Darwin's book were just the most visible manifestation of a very broadly based reassessment of the kinds of assumptions and values on which the earlier generation of Herschel, Whewell, etc., had constructed their view of human knowledge and truth. The reception of non-Euclidean ideas was an integral part of this radical reevaluation of human intellectual powers and aspirations.

NON-EUCLIDEAN GEOMETRY IN 1870

The early nineteenth-century researches of the "synthetic" non-Euclidean geometers, Lobachevskii and Bólyai, concerning Euclid's fifth, or parallel, postulate had led each of them independently to conclude that this controversial postulate was not essential to the development of a consistent geometrical system. Both men believed that this conclusion had important implications for the claim that Euclidean geometry was the exact description of spatial reality. Each of them felt that simply by developing consistent axiom systems which did not rely on Euclid's fifth postulate, he had challenged the exact truth of Euclidean geometry.

The basis for the claim that merely creating an alternative geometrical structure undercut the truth of Euclidean geometry can be seen in Whewell's arguments about the difference between necessary and contingent truth. The mark of geometrical truth, Whewell had claimed, was its necessity: once a geometrical theorem was fully understood, it was impossible to conceive that it could be otherwise. It might require a complicated demonstration to show that the sum of the angles of a triangle is 180°, but when the terms of this theorem were clearly understood, one knew it absolutely. In their work, Lobachevskii and Bólyai claimed to have created alternative geometries—geometrical systems in which triangles *were* otherwise. Their researches changed nothing within the old Euclidean framework. But if they were interpreted as equally valid possibilities for describing space, they severely undercut the kind of truth Euclid's system had been granted on the basis of its uniqueness.

Because they mounted their challenge to Euclid by developing whole new geometrical systems, Lobachevskii and Bólyai have been called "synthetic" geometers. Several decades later Riemann and Helmholtz published research which, they asserted, also led to the conclusion that Euclid's geometry was not necessarily true. These Germans came to their conclusion by a different route, however. Whereas Lobachevskii and Bólyai meticulously constructed geometrical systems which, they maintained, described alternative spaces, Riemann and Helmholtz claimed to have analyzed the classical spatial concept into its component parts. By considering each of these elements in turn, they pointed to various places where alternatives were possible.

It is tempting to label Riemann's and Helmholtz's geometries "analytic" in order to emphasize the contrast with their synthetic predecessors. However, because it was in notions of distance that they found the crucial flexibility which enabled them to create their new geometries, they can equally well be called "metric" geometers. This term points to an important relationship with projective geometry which will be discussed in the next chapter, and for that reason it will be used here.

The major figure who initially brought non-Euclidean ideas to England was Hermann von Helmholtz. Helmholtz is one of the extraordinary universal scientists of the late nineteenth century. His place in the history of physics is secure on the basis of his 1847 work, *Über die Erhaltung der Kraft,* which was critical in the development of thermodynamics. He was originally trained as a physician, however, and throughout his life his most sustained interest was in the physiology of the senses. His three volume work, *Handbuch der physiologischen Optik* (1856–66) is a classic. His *Die Lehre von dem Tonempfindungen,* published in 1863, is equally important to physiological acoustics. Helmholtz was not confined by his physiological interests, however. He followed problems generated in that context in whatever directions they might lead. His work in non-Euclidean geometry was the offshoot of specific problems which fell out of his optical researches.[24]

Helmholtz's non-Euclidean papers contain physiological, philosophical and mathematical arguments entwined so closely as to be virtually indistinguishable from one another. This aspect of his style was often criticized in the increasingly specialized world of continental mathematics.[25] In England, however, it lent power to Helmholtz's ideas. As a thinker, he embodied just the kind of intellectual ideal upheld by the English unitary view of truth. Moreover, he was

---

[24] For a more detailed treatment of the development of Helmholtz's geometrical ideas and their relationship to his other interests, see Joan L. Richards, "The Evolution of Empiricism: Hermann von Helmholtz and the Foundations of Geometry," *The British Journal for the Philosophy of Science,* 28 (1977):135–53.

[25] So, for example, Felix Klein wrote: "Helmholtz ist nicht Mathematiker von Beruf, er ist Physiker, Physiologe. . . . Mit dieser nicht mathematischen Eigenschaft von Helmholtz hängt es dann auch zusammen, dass er den mathematischen Teil seiner Betractungen, nicht mit der Grundlichkeit behandelt und durcharbeitet, wie man es bei einem Mathematiker von Fach fordern wird." *Nicht-Euklidische Geometrie I,* (Göttingen, 1892), p. 260.

a master publicist; following his arguments is still an excellent way to make the work of the metric geometers clear. It is from his work that most English intellectuals gleaned their initial understanding of non-Euclidean geometry.

Helmholtz's first non-Euclidean article in English was the brief "Axioms of Geometry," published in *The Academy*, February 12, 1870. This was the first English summary of the non-Euclidean ideas which had been developed on the continent since the turn of the century. It focused primarily on the metric ideas developed by Riemann and Helmholtz, only touching peripherally on earlier synthetic work. The technical mathematics in the article was scanty; Helmholtz's primary interest lay in presenting the "general drift" of the mathematical investigations. This "drift" can be characterized as epistemological; Helmholtz claimed that the new geometrical research addressed "a question, which . . . may be made generally interesting to all who have studied even the elements of mathematics, and which, at the same time, is immediately connected with the highest problems regarding the nature of the human understanding."[26]

Helmholtz began by comparing the geometrical ideas he assumed would be developed by a group of imaginary two-dimensional creatures living on a plane and a similar group living on the surface of a sphere. The planar group, he maintained, would experience a Euclidean world and would hence develop a Euclidean geometry, albeit limited to two dimensions. Creatures living on the two-dimensional surface of a sphere would perhaps initially agree with the geometry created by plane-dwellers, because for small areas the geometry of a sphere closely approximates that of a plane. As their experience increased, however, this group would recognize that, if followed far enough, every pair of straight lines—the great circles of a sphere—would intersect in two points. This would be true even of lines which might initially seem parallel; the earth's lines of longitude are parallel at the equator but intersect at the poles. In addition, sphere-dwellers would learn that although the angles of small triangles might sum to 180°, those of large triangles would have larger sums; the angles in triangles formed by two longitude lines and the equator, for example, are larger than those of straight-sided

---

[26] Helmholtz, "Axioms," (1870):128.

triangles because the sides bulge to conform to the earth's curvature.

When these facts were taken into account, the sphere-dweller's geometry would be fundamentally different from that of the plane-dwellers. Creatures living on the two-dimensional surface of a sphere would assert that every pair of straight lines intersects in exactly two points. Sphere-dwellers would also know that the sum of the angles of all triangles would be greater than two right angles, and that as the area of the triangle increased, this sum would also increase.

Having thus presented his interpretation of the geometries which would be developed in the two-dimensional case, Helmholtz extended his discussion to include space of three dimensions. In making this extension, he noted:

> These results regarding surfaces or spaces extended in two dimensions only can be illustrated, as we have tried to do, because we live in a space of three dimensions and can represent in our ideas, or model in reality, other surfaces than the plane. . . . When, however, we try to extend these researches to space of three dimensions, the difficulty increases, because we know in reality only space as it exists, and cannot represent even in our ideas any other kind of space. This part of the investigation, therefore, can be carried on only in the abstract way of mathematical analysis.[27]

Since Helmholtz felt that he could not construct conceptual models analogous to the plane and sphere he had introduced for his two-dimensional systems, he resorted to analytical considerations in order to generalize from the relative clarity of the two-dimensional case to the more-difficult-to-picture three-dimensional one. The basic structure on which he constructed his analytic description of three-dimensional space was that of a manifold, an idea he had drawn from Riemann's "Habilitationsvortrag."[28]

A manifold in its most general modern signification is simply a collection of objects or points. A surprising variety of common

---

[27] Ibid., p. 129.

[28] Helmholtz translated Riemann's term *Mannichfaltigkeit* as "variety." Since "manifold" is the English term commonly used which corresponds to Riemann's *Mannichfaltigkeit*, and since "variety" has come to have other connotations, I am here diverging from Helmholtz's usage.

things can be abstractly interpreted through this basic structure. For example, colors can be regarded as points in a manifold, in which each color can be specified by three numbers which show the ratio of the primary colors in the composite. Because exactly three numbers are required to completely specify any color, the color manifold would be three-dimensional. Geometrical space is perhaps the most obvious manifold. Like the color manifold, each of its points can be exactly specified by assigning it three numbers, usually the coordinates of the x, y and z axes. Riemann had noticed that it was the possibility of comparing lengths which set the structure of space apart from other such three-dimensional manifolds. In space, the coordinates of a point are specified by their lengths or distances from the origin; this kind of structure is lacking in a color manifold. Thus the measurement of distance seemed to be the fundamental property which distinguished three-dimensional spaces from the more general category of a three-dimensional manifold. In a spatial manifold, points are completely defined by their positions which are determined by measurements of length.

A critical aspect of Riemann's and Helmholtz's approach to space through manifolds was an emphasis on local properties as a way to generate general spatial knowledge. The challenge for Helmholtz's plane- or sphere-dwellers was not how to describe their world from the outside, embedded in the three-dimensional perspective, but rather how to understand it from within their restricted two-dimensional point of view. The focus of Helmholtz's concern was on these aspects of their limited, local experience from which they could generate an understanding of the overall structure of their environment.

This problem had been approached mathematically in 1827 by Carl Friedrich Gauss in a paper entitled "Disquisitiones Generales Circa Superficies Curvas." Gauss there defined a measure of curvature which was intrinsic to a small piece of a curve or a surface.[29]

---

[29] Carolo Friderico Gauss, "Disquisitiones Generales Circa Superficies Curvas," *Commentationes societatis regiae scientiarum Gottingensis recentiores*, 6 (1828), reprinted in *Carl Friedrich Gauss Werke*, 12 vols. (Königlichen Gesellschaft der Wissenshaften zu Göttingen, 1873), 4, pp. 217–58.

With this measure, one could find the tightness of a curve or surface; that is, one could determine the radius of the circle or ball on which it would fit exactly, without having to step back to see the whole. More concretely, one could say that Gauss developed a way to measure the difference in the curvatures of the edges of a penny and a half-dollar, without placing one on the other. Using Gauss's approach, local studies of curves were adequate to generate knowledge of the curvature of the surfaces on which they were drawn. This was essentially the approach that allowed Helmholtz's beings to recognize the differences between planar and spherical experiences, or even spheres of different radii, without stepping outside of their two-dimensional experience.

Gauss had confined his remarks to curves traced on two dimensional surfaces, but in his "Habilitationsvortrag," Riemann extended Gauss's method of locally defining two-dimensional curvature to the three-dimensional case. Since three-dimensional manifolds have the same number of dimensions as everyday space, we cannot step outside of them in order to view them locally in the same way we can two-dimensional manifolds. We *can*, however, consider how they act locally, that is, what properties distance measurements would display in them. From these local properties we can construct an overall picture of what they are like in the same way that Helmholtz's two-dimensional sphere-dwellers could construct an understanding of the surface of their sphere. In this way, Riemann developed mathematical descriptions of three-dimensional spaces which were not flat like Euclidean space. His spaces could have either a "positive curvature" in which case they were curved analogously to a beach ball, "zero curvature" in which case they were like a flat plane or "negative curvature" which is harder to model than the other two cases but locally curves rather like a saddle.[30]

Riemann's spaces of negative curvature were geometrically the

---

[30] A model for pseudospherical space was first proposed by E. Beltrami in "Saggio di interpretazione della geometria non-euclidea," *Giornale di matematiche*, 6 (1868):284–312. This was translated by J. Houël as "Essai d'Interpretation de la Géométrie non-Euclidienne," *Annales Scientifiques de l'Ecole Normale Supérieur*, 6 (1869):251–88.

same as the spaces Bólyai and Lobachevskii had derived from the hypothesis of the acute angle. His spaces of positive curvature represented a new departure, however. They correspond to those spaces which would be generated from the hypothesis of the obtuse angle. All non-Euclidean investigators since Saccheri had dismissed such geometries because of arguments about the infinity of space. This obtuse angle hypothesis would generate a geometry like that on a sphere where lines are not infinite but rather turn back upon themselves. A series of arguments dating back to Aristotle, however, made it clear that space must be infinite. If it were not, one argument ran, what would happen at its edges? What would prevent someone from standing at the edge and extending his arm into the void beyond space?

Riemann's spaces of positive curvature, which corresponded to the hypothesis of the obtuse angle, rested on a different approach to this dilemma. He drew a distinction between boundless and infinite. The surface of a beach ball might not be infinite in the way a plane is, but it is nonetheless boundless. Wherever one chooses to be on a sphere one can still move freely in any direction; there is no edge beyond which one cannot extend an arm. Recognizing this allowed Riemann to reject disproofs of the hypothesis of the obtuse angle which had been based on the assumption that space was infinite. He proposed that spaces of positive curvature, which were boundless though finite, were on an essentially equal footing as those with negative or zero curvatures.

In his "Habilitationsvortrag," Riemann did not limit his analysis to spaces with a constant curvature. Such spaces would be analogous to polished planes or smooth beach balls, where the amount of curvature would be the same wherever one was. He pointed out that spaces might equally well have nonconstant curvature and be analogous to bumpy or irregularly twisted surfaces; one might equally well talk of pitted planes, oddly curved rugby balls or even footballs with pointed ends. He even suggested in his conclusion that our physical space might be irregularly curved. In our everyday experience, space appears to have a constant curvature which, furthermore, appears to be very close to zero. In the realm of the infinitely large or small, however, it might be irregularly twisted, its smoothness marred by pits and lumps. Such a state of affairs, Riemann hypothesized, might account for puzzling phenomena like electric and magnetic effects on the molecular level.[31]

All of these ideas—that spaces were three-dimensional manifolds characterized by properties of distance measurements which gave them positive, negative or zero curvature—were packed into Riemann's incredibly dense "Habilitationsvortrag." It is not surprising that the depth of implication lurking in the brief compass of this paper made it inscrutable to most of his contemporaries. When they wished to understand Riemann's ideas they tended to consult Helmholtz who relied on Riemann but wrote much more clearly. Helmholtz's presentation, however, significantly affected the picture he painted.

Riemann had been primarily interested in uncovering the kinds of logical possibilities which were inherent in the notion of space and then looking at the physical experiences which might lead us to choose one such notion over another. Helmholtz was also interested in analyzing the different spatial possibilities which were logically possible. For him, however, the issue was to identify the basic experiences all human beings shared which led them consistently to construct a Euclidean notion of space. He was primarily a physiologist and psychologist wishing to understand those experiences which led human beings to understand space as they did.

In his efforts, Helmholtz focused on the motions of rigid bodies, which he believed provided the critical experience on which our notion of space rested. Any child—even a blind one—learns early that objects can be moved around in space without being deformed; a rattle does not change its size or shape as it is moved around a crib. Helmholtz believed that this experientially determined fact was critical in determining our notion of space. From it he derived the measurement of distance which Riemann had taken as a given for defining space.

---

[31] Riemann's suggestion that space need not be considered to be of constant curvature required a radical change in the physical perception of space. For discussion of the nature of the change required, see Max Jammer, *Concepts of Space: The History of Theories of Space in Physics*, with a Foreword by Albert Einstein, (Cambridge: Harvard University Press, 1954), Ch 5. This idea was briefly explored in William K. Clifford, "On the Space-Theory of Matter" (Read Feb. 21, 1870), *Proceedings of the Cambridge Philosophical Society*, 2 (1876):157–58.

Helmholtz's assumption that rigid body motion without deformation was possible is equivalent to the assumption of constant curvature. Objects can be slid along smooth surfaces without being distorted, but on rough or irregularly twisted ones, their forms will change with their position; this can be seen if one projects a picture onto a rough surface. Thus, by considering Riemann's work from his physiological point of view, Helmholtz significantly narrowed the spatial possibilities Riemann had been willing to consider. Even so, Helmholtz's approach allowed the possibility of three different forms of space: those with constant positive, negative or zero curvature. His analysis of infantile experience with rigid bodies led to the conclusion that Euclidean geometry was not uniquely determined.

Helmholtz interpreted non-Euclidean geometries through an intellectual perspective which was particularly powerful in England where geometry was self-consciously embedded in an over-arching epistemological tapestry. He focused directly on the question of what we know about space and how we know it. For him, non-Euclidean geometry was the result of analytical investigations of issues of perceptual interest, with strong implications for the *real* space of human experience and conception. He explicitly stated some of these implications at the end of his first English article:

> We may resume [sum up] the results of these investigations by saying, that the axioms on which our geometrical system is based, are not necessary truths, depending solely in irrefragable laws of our thinking. On the contrary, other systems of geometry may be developed analytically with perfect logical consistency. Our axioms are, indeed, the scientific expression of a most general fact of experience, the fact, namely, that in our space bodies can move freely without altering their form. From this fact of experience it follows, that our space is a space of constant curvature, but the value of this curvature can be found only by actual measurements.[32]

That geometry was *not* the study of necessary truth, as Whewell had defined it, was a conclusion which could be drawn directly from Helmholtz's interpretation and development of Riemann's mathematical results. The physiologist tied Riemann's highly abstract ideas

---

[32] Helmholtz, "Axioms," (1870):130.

tightly to the most common set of experiences and argued that these experiences were the basis for the notion of space. This foundation, Helmholtz emphasized, was inadequate to fix our notions as Euclidean. As a result, claims for a special geometrical truth status based on Euclid's uniqueness were fundamentally flawed.

Helmholtz's analysis of the spatial concept, on the surface so clear and persuasive, raised a wide variety of issues. He presented a powerful philosophical and epistemological message as an integral part of his non-Euclidean geometry. A central part of this message was empirical; his whole geometrical analysis was presaged on the assumption that knowledge of space was formed through experience. Implicit in this assumption were a number of issues which engendered immediate response among Helmholtz's English readers. Most generally the issues can be characterized as turning on notions of truth; William Stanley Jevons engaged Helmholtz on this level in an almost immediate response to the 1870 article. More narrowly focused issues were raised in the years which followed. In 1873, Riemann's "Habilitationsvortrag" was published in English. His work was more narrowly mathematical than Helmholtz's and rekindled the internal mathematical debate over the relationship of analysis to geometry, this time couched as the question of how analytically constructed manifolds could be said to relate to the geometrical concept of space. In 1876, Helmholtz again published in English, a longer paper this time, which raised yet more detailed questions of how people think and perceive. This sparked yet another discussion which focused more specifically on philosophical and epistemological issues in the geometrical context. Thus, in the decade and a half following Helmholtz's first brief article about non-Euclidean geometry, English discussions of the subject spread in many directions. It seemed that some kind of irresistible force was being applied to Euclid's immovable geometry, and many were anxious about what the outcome would be.

THE IMMEDIATE ENGLISH RESPONSE TO NON-EUCLIDEAN GEOMETRY

The Nature of Geometrical Truth: Helmholtz and Jevons

In 1871, the year after Helmholtz published "The Axioms of Geometry," the logician and economist William Stanley Jevons responded in *Nature*. The impact of Helmholtz's article can be judged by the forum of reply; *Nature* was and is a journal with a large, educated readership that extended well beyond the confines of the mathematical community. Jevons framed the issue raised by Helmholtz's article as follows:

> The opinions set forth by him [Helmholtz] were based upon the latest speculations of German geometers, so that a new light seemed to be thrown upon a subject which has long been a cause of ceaseless controversy. While one party of philosophers, especially Kant and the great German school, have pointed to the certainty of geometrical axioms as a proof that these truths must be derived from the conditions of the thinking mind, another party hold that they are empirical and derived, like other laws of nature, from observation and induction.[33]

Thus, Jevons's initial response was to Helmholtz as an empirical learning theorist who argued that human intellectual concepts are drawn from experience.

Although this was his starting point, Jevons quickly moved out of developmental questions to focus on the philosophical question of the nature of geometrical truth. He saw Helmholtz as providing evidence for an empirical interpretation of this truth by asserting that under different conditions, intelligent beings would construct different geometries. Jevons opened his discussion of Helmholtz's ideas with the following statement about the truth of geometric axioms:

> All these geometrical exercises have no bearing whatever upon the philosophical questions in dispute. Euclid's elements would be neither more nor less true in one such world than another; they would be only more or less applicable. Even in a world where the figures of plane geometry could not exist, the principles of plane geometry might have been developed by intellects such as some men have possessed. And if, in the course of time, the curvature of our space should be detected, it will not falsify our geometry, but merely necessitate the extension of our books upon the subject.[34]

---

[33] W[illiam] S[tanley] Jevons, "Helmholtz on the Axioms of Geometry," *Nature*, 4 (1871):481.
[34] Ibid.

Jevons's position was, essentially, that although Helmholtz had been successful in creating models of a world in which the Euclidean axioms of geometry would not accurately describe experience, the existence of these models did not affect the *truth* of Euclid's axioms; it was simply the case that within the peculiar situations Helmholtz had hypothesized, Euclid's axioms would be less *applicable*.

Jevons's distinction between truth and applicability suggested the independence of geometric truth from experiential reality. It did not, however, necessarily imply that geometry had no real spatial referent. He did not conclude that geometrical systems are developed abstractly, independent of a spatial reality. On the contrary, the rest of Jevons's article clearly shows that in his view geometrical systems were true descriptions of a reality yet more fundamental than that which could be derived from experience. Geometrical truth was the transcendental truth about the essential reality lying behind mere perceptual experience. The bulk of Jevons's article was directed towards demonstrating the ways in which Helmholtz's two-dimensional sphere-dwellers would come to know this truth. His goal was to establish that if Helmholtz's two-dimensional creatures possessed a human intelligence, they would also possess the critical ability to come to real knowledge, in this case that space was both Euclidean and of three dimensions.

The crucial geometrical fact upon which Jevons based his argument came from the world of the infinitesimally small. The essential flaw he detected in Helmholtz's argument lay in the fact that in both spherical and pseudospherical spaces, all geometrical relations reduced to Euclidean relations in the infinitely small. This fact, he maintained, would ultimately lead dwellers in both spaces to know the truly Euclidean nature of space, even though their experience of it was distorted.

> The whole of plane geometry would be as true to them [two-dimensional sphere- or pseudosphere-dwellers] as to us, except that it would only be exactly true of infinitely small figures. The principles of the subject would certainly be no more difficult than those of the Differential Calculus, so that if a Euclid could not, at least an Archimedes, a Newton, or a Leibnitz of the spherical world would certainly have composed the books of Euclid, much as we have them. Nay, provided that their figures were drawn sufficiently small, they

could verify all truths concerning straight lines just as closely as we can.[35]

Jevons did not merely indicate that the two-dimensional inhabitants of a surface with nonzero curvature *could* construct a plane geometry like that of Euclid. He argued that they *would* see that it contained the real truth behind their distorted experience. He wrote,

> I do not think that the geometers of the spherical world would be under any greater difficulties than our mathematicians are in developing a science of mechanics, which is generally true only of infinitesimals. Similarly in all the other supposed universes plane geometry would be approximately true in fact, and exactly true in theory, which is all we can say of this universe.[36]

Jevons argued this point not only with respect to the curvature of these dwellers' experience but also with respect to its dimensions. Even though they would have no direct perception of a third dimension, a careful analysis of their existence would reveal it to them:

> They could not fail to be struck with the fact that their geometry of finite figures differed from that of infinitesimals, and an analysis of this mysterious difference would certainly lead them to all the properties of tridimensional space.[37]

Even though Jevons accepted the two-dimensional models Helmholtz created, he did not grant a distinction between the geometrical truth which would be developed in such worlds and that created in a three-dimensional Euclidean one. The inhabitants of a non-Euclidean world would perhaps experience some difficulty in their attempts to understand the truth, but in the long run they would grasp the ultimate reality of the three-dimensional Euclidean space in which their distorted world was embedded.

Helmholtz replied to Jevons in *The Academy,* February 1, 1872. His remarks were directed squarely at Jevons's interpretation of

---

[35] Ibid.
[36] Ibid.
[37] Ibid., p. 482.

truth. He pointed to two alternative meanings for truth—experiential truth and mathematical truth. Helmholtz defined experiential truth as follows:

> Where I say that geometrical axioms are true or not true for beings living in a space of a certain description, I mean that they are true or not true in relation to those points, or lines, or surfaces, which can be constructed in these spaces, and which can become objects of real perception to those beings.[38]

This truth he distinguished from truth which could be generated analytically. Helmholtz gave as an example analytical ideas of spaces of more than three dimensions.[39] These entities, Helmholtz argued, were true in that they generated consistent results. This consistency did not imply ontological validity, however. "But for all this," Helmholtz wrote,

> no mathematician ever came to the conclusion that a fourth dimension of space exists, even though he find it convenient to write his equations as if it existed. And I cannot see why the mathematical intellects of a spherical or pseudo-spherical world should come to another conclusion. . . . Points and lines in such a space would have no more meaning to them than length in the direction of the fourth co-ordinate can have for us, although we introduce such a co-ordinate into our calculations.[40]

The discussion between Jevons and Helmholtz clearly locates the point of strain created within the nineteenth-century intellectual fabric by the introduction of non-Euclidean geometry; this was the nature of the truth developed in geometry. For Helmholtz, there was an experiential truth which was generated to describe exactly the relations perceived only inexactly among the objects of experience. This truth grew out of the interaction of intellect and experience, and hence required both. It was distinct from mathematical

---

[38] Hermann von Helmholtz, "The Axioms of Geometry," *The Academy*, 3 (1872):52.
[39] He also pointed to the imaginary points of intersection of circles, which will be discussed in the next chapter.
[40] Ibid., pp. 52–53.

truth because mathematical truth could be subjectively generated from axioms. This kind of truth required intellect but not experience. For Helmholtz, who was an empirical scientist, the first variety of truth, experiential truth, was of primary interest. Purely subjective mathematical truth held no ontological implications, and hence did not interest him.

Jevons's approach was markedly different. Educated within the English descriptive tradition of mathematics, he did not accept Helmholtz's view that mathematical truth was empty, nor the assertion that the relationship between mathematics and reality was merely experiential. Instead he focused on a third kind of truth, a transcendental or necessary truth, of which geometry was the exemplar. This was a truth in which experience and intellect were joined. It was not bounded by experience, however; it was recognized rather than learned. The mark of this truth was not Helmholtz's criterion of applicability—that criterion was essentially experiential and irrelevant. Rather it was conceptual clarity which characterized mathematical truth—a clarity which could be generated and understood through sustained study.

Jevons's judgment of non-Euclidean geometries in terms of an essentially Whewellian category of necessary, mathematical truth—which had a primary ontological status independent of its practical applicability—was not peculiar to him. It was a position which was widely accepted within the British mathematical community in the years that followed. Echoes of it can be found, for example, in Arthur Cayley's 1883 Presidential Address to the British Association for the Advancement of Science. Here Cayley hinted at the same criteria when he considered the geometry of the sphere-dwellers. He noted that a spherical geometry would "accurately represent" the realities of the given situation, but hastened to note that the Euclidean geometry the sphere-dweller would construct from infinitesimals would be "a true system" applying to "an ideal space, not the space of their experience."[41]

The discussion between Helmholtz and Jevons suggests the basic parameters of the issues non-Euclidean ideas raised in the English

---

[41] Arthur Cayley, "Presidential Address," *Report of the Fifty-third Meeting of the BAAS held at Southport in August 1883*, (London: John Murray, 1884), p. 10.

context. It was, however, a very brief and necessarily general exchange. Some of the specific details of the English position were argued in subsequent interchanges. The mathematical implications, involving the relationship between analysis and geometry, are clarified in the writings of Riemann and two English mathematicians, William Clifford and Samuel Roberts. The philosophical implications were pursued in greater detail in a later discussion involving Helmholtz who, this time, was defending his case against J. P. N. Land.

Although the specific issues in the mathematical and philosophical discussions were often rather different, the central problem was essentially the same for both of them. In the English formulation of the situation, the mark of geometry's necessary truth was the impossibility of conceiving it could be otherwise. This criterion seemed clear when no alternative geometries were at hand; their very absence seemed proof that they were not possible. With non-Euclidean developments, however, the difficulty of settling the essentially subjective question of what could be "conceived" became evident. The confusion inherent in deciding who could conceive what, which appeared in the discussions surrounding higher dimensions, arose again with respect to non-Euclidean geometry. In the non-Euclidean discussions, the nature of conceivability was directly scrutinized, both mathematically and philosophically.

Analyzing the Concept of Space: Riemann, Clifford and Roberts

An important facet of the challenge which Helmholtz raised for the necessary truth of geometry involved the relationship between analysis and geometry. In the last chapter this problem was considered in the discussions surrounding Gregory's imaginary-branched curves and those about multidimensional spaces. In both of these cases, the primacy of geometrical conception was firmly defended against the encroachments of analytic development. Metric non-Euclidean geometry raised the issue again, this time in an even starker form. The metric geometers claimed to have dissected the concept of space into its component parts. They did not merely claim that their analysis suggested places where geometrical ideas, like those of plane curves or dimensionality, ought to be reexamined. They claimed to have reduced geometry to its analytical essence so completely that their analytical construction would replace the conception of space which had defined geometry for so long.

The opening paragraphs of Riemann's "Habilitationsvortrag" clearly indicate this point of view. Here he unequivocally rejected the prevailing interpretation of space as an essentially amorphous concept which could be mathematically described but not analytically defined.

> It is known that geometry assumes, as things given, both the notion of space and the first principles of constructions in space. She gives definitions of them which are merely nominal, while the true determinations appear in the form of axioms. The relation of these assumptions remains consequently in darkness; we neither perceive whether and how far their connection is necessary, nor, *a priori*, whether it is possible.
>
> From Euclid to Legendre (to name the most famous of modern reforming geometers) this darkness was cleared up neither by mathematicians nor by such philosophers as concerned themselves with it. The reason of this is doubtless that the general notion of multiply extended magnitudes (in which space-magnitudes are included) remained entirely unworked.[42]

Riemann's goal of dispelling the "darkness" surrounding the relations among spatial assumptions led him to his study of multidimensional magnitudes. The basic premise of his study was that spatial magnitude was no more than a particular case of this general analytical construct. Before any conclusion about the nature of space could be drawn from his work, one had to accept this fundamental assumption.

This point, implicit in Riemann's terse and complex prose, was presented to English audiences in a more accessible form by the mathematician William Clifford, who acted as the major English spokesman for Riemann's ideas. In addition to making Riemann's work directly available by translating it for *Nature*, Clifford inter-

---

[42] Riemann, "Hypotheses," (1873):14.

preted it in a series of lectures entitled "The Philosophy of the Pure Sciences," which he delivered at the Royal Institution in 1874 and subsequently published in the *Contemporary Review*. In the second of these lectures, subtitled "The Postulates of the Science of Space," Clifford carefully spelled out the implications he found in Riemann's new approach to geometry.

Clifford argued that Riemann's analysis of the spatial concept into its defining components had destroyed the illusion of certainty on which geometry's epistemological claims had rested. He wrote:

> It was Riemann, however, who first accomplished the task of analyzing all the assumptions of geometry, and shewing which of them were independent. This very disentangling and separation of them is sufficient to deprive them for the geometer of their exactness and necessity; for the process by which it is effected consists in shewing the possibility of conceiving these suppositions one by one to be untrue; whereby it is clearly made out how much is supposed.[43]

The reason Riemann's analysis had destroyed geometry's necessary truth was that it had pinpointed exactly the issues before clumped inexactly in the conviction that non-Euclidean alternatives could not be conceived. Clifford's speech was a masterful exercise in disentangling the assumptions of geometry which Riemann's analysis had identified and discussing alternative constructions. Clifford identified four spatial postulates: those of continuity, of elementary flatness, of superposition and of similarity. The notion of elementary flatness he juxtaposed against the curved notions suggested by Helmholtz's sphere-dwellers. That of superposition corresponded basically to Helmholtz's notion of rigid bodies which could be moved from place to place without deformation. Simply being able to think of alternatives to these assumptions, Clifford argued, was adequate to explode the celebrated claim that alternatives to Euclid's geometry were inconceivable.

Despite the charm and clarity of his presentation, Clifford's view of non-Euclidean geometry and its implications was not widely

---

[43] Clifford, "Postulates" (1875):374–75.

accepted. A clear expression of an alternative response to the kind of spatial analysis claimed by the metric geometers is in a speech given in 1882 by Samuel Roberts, the president of the London Mathematical Society. Roberts clearly indicated his concern with the epistemological issues raised by the new geometrical approach in the title of his address: "Remarks on Mathematical Terminology, and the Philosophic Bearing of Recent Mathematical Speculations Concerning the Realities of Space."

Roberts first acknowledged that geometrical terminology could serve a useful analogical function in clarifying analytical conceptions. However, he was quick to point out the perils of thus extending geometrical language as if the extensions implied reality. With respect to non-Euclidean geometry, he insisted that "the progress of analysis has in no way affected the philosophical status of our notions of space and time, and that, consequently, metaphysicians need feel no alarm at what is going on behind the veil of mathematical symbolism."[44]

Roberts's categorical denial of the philosophical significance of the new geometrical researches was possible because he did not accept the metric claims to have defined the concept of space through their analysis. Addressing himself to Riemann's work, he noted,

> The multiply extended magnitudes of which he [Riemann] speaks are quantitative conceptions, after all. In certain cases, we can fit them to real cognitions. They are not, in that case, the cognitions, but an expression of certain quantitative relations to which the cognitions conform.[45]

He elaborated on this denial, writing:

> if realities exist which may be adequately interpreted in these quantitative laws by means of a triply-extended magnitude of constant quasi-curvature, there is no reason to affirm that those cognitions are

---

[44] Samuel Roberts, "Remarks on Mathematical Terminology, and the Philosophic Bearing of Recent Mathematical Speculations Concerning the Realities of Space," (Read Nov. 9, 1882), *Proceedings of the London Mathematical Society,* 14 (1882–83):11.
[45] Ibid., p. 12.

spatial in their character. How can we legitimately assume this, when we ourselves have different notions forming one-fold aggregates with similar measure relations, notions so different, yet so wonderfully connected, as time and linear space?[46]

In thus rejecting the metric claim to have analyzed the spatial concept to its defining components, Roberts was able to preserve the conceptual approach to space as primary. He preferred leaving the relations among spatial axioms in obscurity to accepting Riemann's analytical attempt to bring the basic elements of the spatial concept to light. In his view, geometry remained the exploration of an essentially undefinable portion of truth.

> The conception of space is admittedly vague and inadequate, until sustained reflection is brought to bear upon it. . . . This being so, it cannot be taken for granted that the most cultured person adequately conceives space, as it confronts the human mind, much less as it confronts minds of a higher order, or as it is in itself. Hence, it is quite to be expected that we should be able from time to time to analyse our spatial notions with greater precision. In this way, the researches of Lobatchewsky are extremely valuable, especially with regard to the theory of parallels.[47]

Roberts's position is a somewhat familiar one. His linguistic emphasis is reminiscent of George Salmon's or William Spottiswoode's approach to higher dimensional spaces. Particularly when filtered through Clifford's persuasive prose, however, Riemann's ideas were hard to dismiss as merely linguistic conventions. Clifford's careful explanations made it difficult to insist that the analytical developments had no spatial cogency, that, like Spottiswoode's rainbows, they would "fade from the scene."[48] This difficulty became even greater after 1876, because then Helmholtz published another English paper, in which he tried not just to analytically

---

[46] Ibid.
[47] Ibid., p. 13.
[48] William Spottiswoode, "Presidential Address," *Report of the Forty-eighth Meeting of the BAAS Held at Dublin in August 1878*, (London: John Murray, 1879), p. 23. For a treatment of the full quotation from which this reference comes, see the final section of chapter one.

structure, but also to experientially describe, non-Euclidean spaces. By doing this he focused attention directly on the question of what it meant to "conceive" an alternative to Euclidean space. This issue was explicitly considered in an interchange generated by Helmholtz's 1876 paper.

Conceivability and Geometry: Helmholtz and J. P. N. Land

In the discussion which surrounded his 1870 paper, Helmholtz had asserted the existence of experimental truth (the contingent truth of the empirical sciences) and of mathematical truth (the truth of mathematical consistency) but he did not explicitly discuss the necessary, transcendental truth which Jevons and other Englishmen had affirmed for geometry. The disagreement between Clifford (or Riemann) and Roberts also did not take the form of an active debate about the concept of space; in fact, Clifford had been dead for three years by the time of Roberts's remarks. However, the relationship between the central characteristic of the spatial concept which marked its necessary truth—conceivability—and the analytic researches of the metric geometers was explicitly raised in another series of English articles, this time published in the new psychological journal *Mind*. Here, in 1876, Helmholtz published a long article which embroiled him in a discussion with the Dutch philosopher J. P. N. Land. Although neither of the discussants in this case was English, *Mind* was an English forum, and the issues raised here had a lasting effect on English geometrical ideas through the rest of the century.

In his 1876 article, which was an English translation of a lecture he had delivered in Heidelberg in 1870, Helmholtz addressed the issue of geometrical truth directly by attempting to establish the conceivability of the alternative geometrical systems Riemann had developed. He did not merely offer alternative axioms, as Clifford had done, but tried to establish that the alternatives were equally possible psychologically. He tried to show that the alternative spaces could be conceived as clearly as Euclidean space. Helmholtz's claims to have established not only the analytical consistency but also the conceivability of the alternative spaces raised serious issues for those who wished to deny the relevance of the new analytical developments to time-honored perceptions of geometrical truth.

Helmholtz opened his discussion by restating the mathematical conclusions of his original *Academy* article. Having shown that spaces of positive or negative curvature could be generated from basic manifolds as easily as could spaces of zero curvature, he elaborated the implications he attached to these mathematical results. He maintained that since the properties of Euclidean space that led to the classical theory of parallels were not implicit in the basic spatial manifold, they could not be "necessities of thought." Therefore he proposed examining the alternative assumption, that they were induced from facts of experience. The key to this alternative approach, Helmholtz claimed, lay in constructing descriptions of the experience one would have in alternative, non-Euclidean spaces. If such constructions were possible, the uniqueness of Euclidean space would be undermined and its necessary truth disproved. He wrote:

> We have now to seek an explanation of the special characteristics of our own flat space, since it appears that they are not implied in the general notion of an extended quantity of three dimensions and of the free mobility of bounded figures therein. Necessities of thought, involved in such a conception, they are not. Let us then examine the opposite assumption as to their origin being empirical, and see if they can be inferred from facts of experience and so established, or if, when tested by experience, they are perhaps to be rejected. If they are of empirical origin we must be able to represent to ourselves connected series of facts indicating a different value for the measure of curvature from that of Euclid's flat space. But if we can imagine such spaces of other sorts, it cannot be maintained that the axioms of geometry are necessary consequences of an *à priori* transcendental form of intuition, as Kant thought.[49]

Helmholtz here argued that the possibility of generating non-Euclidean spaces from the fundamental manifold of three dimensions raised the issue of whether such spaces were imaginable (or conceivable). He felt that this depended upon whether one could describe the experiences one would have within a non-Euclidean space.

---

[49] Hermann von Helmholtz, "The Origin and Meaning of Geometrical Axioms," *Mind*, 1 (1876):314.

For this reason, Helmholtz attempted to represent what it would feel like to be in a non-Euclidean world. He likened existence in these worlds to existence in the world one could observe behind a curved mirror or through prism glasses and carefully detailed what experience would seem like in such situations. Thus, for example, he described the experience of a person raised in a Euclidean world, but moved to a "pseudospherical one"—one with negative curvature where the acute angle hypothesis would hold—as follows:

> He would think he saw the most remote objects round about him at a finite distance, let us suppose a hundred feet off. But as he approached these distant objects, they would dilate before him, though more in the third dimension than superficially, while behind him they would contract. He would know that his eye judged wrongly. If he saw two straight lines which in his estimate ran parallel for the hundred feet to his world's end, he would find on following them that the farther he advanced the more they diverged, because of the dilatation of all the objects to which he approached. On the other hand behind him their distance would seem to diminish, so that as he advanced they would appear always to diverge more and more. But two straight lines which from his first position seemed to converge to one and the same point of the background a hundred feet distant, would continue to do this however far he went, and he would never reach their point of intersection.[50]

The elaborate care with which Helmholtz described these non-Euclidean experiences was not merely the product of fertile fancy. It was directed towards a definite end. He hoped thereby to establish that non-Euclidean spaces could be represented just as easily as and hence were cognitive alternatives to Euclidean ones. He felt that success in this venture, which would allow him to conclude that Euclidean geometry was not the expression of necessary truth, depended on his being able to construct the sensible experiences one would have in a non-Euclidean space.

Helmholtz's experiential representations of non-Euclidean spaces evoked an immediate response. In his article, "Kant's Space and Modern Mathematics," published in *Mind* a year after Helmholtz's work appeared, J. P. N. Land challenged the basic assumption which

---

[50] Ibid., p. 317.

would make Helmholtz's elaborate descriptions at all relevant to the issue of spatial conception. He denied that describing experiences in these spaces was sufficient to make them conceivable. By rejecting Helmholtz's primary operating principle, that our conception of space is determined by our experience of it, Land argued for the transcendental truth of Euclidean space even in the face of Helmholtz's experientially described conceivability models.

To Helmholtz, the ability to describe a situation and describe the series of sensations we would have were we in that situation constituted the ability to imagine. Land criticized Helmholtz for this use of the term, writing,

> In the present case, the first question is whether any sort of space besides the space of Euclid be capable of being imagined. More than three dimensions, it is allowed, we are quite unable to represent. But we are told of spherical and pseudospherical space, and non-Euclideans exert all their powers to legitimate these as space by making them imaginable. We do not find that they succeed in this, unless the notion of imaginability be stretched far beyond what Kantians and others understand by the word.[51]

In the first place, like Roberts, Land did not accept that the analytical description of non-Euclidean spaces was sufficient to render them imaginable.

> And when we are assured that Beltrami has rendered relations in pseudospherical space of three dimensions imaginable by a process which substitutes straight lines for curves, planes for curved surfaces, and points on the surface of a finite sphere for infinitely distant points, we might as well believe that a cone is rendered sufficiently imaginable to a pupil by merely showing its projection upon a plane as a circle or a triangle.[52]

Land went further, however. He rejected not only Helmholtz's analytical models but also his more concrete descriptions of experience behind circus mirrors or through prism glasses as being irrelevant to the conceivability of these non-Euclidean worlds. To do

---

[51] J. P. N. Land, "Kant's Space and Modern Mathematics," *Mind,* 2 (1877):41.

[52] Ibid., pp. 41–42.

this, he developed an interesting twist in the argument on the relationship between cognitive and empirical reality. Land pointed out that the world which we actually perceive does not itself correspond to Euclidean space. Instead we have learned to accommodate this non-Euclidean experience to Euclidean reality through practice. He wrote,

> As to the image in a convex mirror, referred to by Prof. Helmholtz in his article, we do not mentally contrast it with our objective world in Euclidean space, but only with the habitual aspect of that world as seen from a given point of view. In the latter also things appear to contract as they retire to a distance. Only we have learned to conceive the objective space as one in which we ourselves are able to move in all directions and shift our point of view at pleasure. So with some practice we actually see those things not growing smaller, but moving away from the place where we may happen to be.[53]

For Land, it followed that Helmholtz's alternatives to three-dimensional, non-Euclidean worlds were as irrelevant to the reality of Euclidean space as was the space of our visual perception.

> The world in the mirror offers itself as a novel aspect of the same world, needing a larger amount of practice for its interpretation, because complicated by unwonted circumstances. As a form of the objective world, which remains the same from whatever point we inspect it, we can imagine, not any space in which motion implies flattening or change of form of any kind, but only the space known from our sense-experience, the space of Euclid. All other 'space' contrived by human ingenuity may be an aggregate with fictitious properties and a consistent algebraical analysis of its own, but space it is called only by courtesy.[54]

Land argued that, given the proper amount of practice, Helmholtz's alternative worlds would be interpreted through Euclidean geometry by their inhabitants. Hence these descriptions did not have significant implications for the essential issue of the imagina-

---

[53] Ibid., p. 42.
[54] Ibid.

bility or conceivability of non-Euclidean spaces. Land held that there was a universal, Euclidean faculty of spatial conception which was the foundation of all human perceptions of space. It was this concept which informed human perceptions of the external world and formed the subject matter of geometry. A world wherein one's experiences were different would simply require that a different set of experiences be integrated with and perceived through that conceptual context. This position allowed continued adherence to a transcendentally true geometry, despite non-Euclidean development.

Land's interpretation in which Helmholtz's conceivability models merely represented different perceptions of essentially the same spatial concept was not a new one. His orientation was commensurate with his contemporaries' interpretations of ideas like those presented in Reid's geometry. Looking again at Roberts's 1882 address, for example, Reid's argument is treated essentially as Land had treated Helmholtz's conceivability models.

> But it is right to remember that the possibility of geometries proceeding from postulates different from those of Euclid, or their equivalents, has long been recognized by philosophers. Berkeley led the way, by distinguishing between visible and tangible magnitude, which, indeed, he regarded as distinct in nature, and only related as sign to thing signified.
>
> Reid's section on the 'Geometry of Visibles,'* with his entertaining account of the 'Idomenians,' who only perceived visible extension, is interesting, because he gives a specimen of the Geometry. . . .
>
> The example suffices, if any were required, to show that our plane geometry is not the only consistent one. I do not see that, philosophically speaking, the parallel-theory of Lobatchewsky, and his 'Géométrie Imaginaire,' alter our position as to the space idea, except in so far as the elaboration of his work is calculated to impress the mind more forcibly.
>
> *"Reid's Collected Writings," Hamilton's Edition, Vol. I, p. 147.[55]

Although they rarely addressed the issue directly, it seems that most English mathematicians in the 1870s dismissed the signficance of

---

[55] Roberts, "Remarks," (1882–3):13.

Helmholtz's conceivability models in this manner. Rather than perceiving them as alternative spaces, they were interpreted as merely
distorted descriptions of a Euclidean spatial reality.

The orientations of Land and Helmholtz were so fundamentally
different that it was very difficult to argue one point directly against
the other. Helmholtz constructed his analysis using two kinds of
truth, mathematical truth based on consistency and empirical truth
based on events in the environment. Neither of these was the essential issue for his opponents who focused on yet a third kind of truth,
which was somewhere between the mathematical and the empirical.
A particular English tradition of interpreting the differences
between empirical and geometrical reality led to ascribing necessary
or transcendental truth to Euclid's geometrical theorems.

In a final English article, "The Origin and Meaning of Geometrical Axioms (II)," Helmholtz attempted to meet this interpretation
head on and to weaken the conceptual stance by exposing it as
meaningless and superfluous. He accepted the distinction between
experientially described, empirical truth and subjectively conceived,
necessary truth for the sake of argument. He then tried to show that
such a distinction was essentially trivial.

> If . . . it is still assumed hypothetically that the axioms are really given
> *a priori* as laws of our space-intuitions, two kinds of equivalence of
> space-magnitudes must be distinguished: a) *Subjective equality* given
> by the hypothetical transcendental intuition; b) *Objective equivalence*
> of the real substrata of space-relations, proved by the equality of
> physical states or actions, existing or going on in what appear to us
> as congruent parts of space. The coincidence of the second with the
> first could be proved only by experience; and as the second would
> alone concern us in our scientific or practical dealings with the objec-
> tive world, the first, in case of discrepancy, must be discounted as a
> *false show.*[56]

Helmholtz's essential objection was to positing barriers that
would stand between human knowing and experiential reality. In

---

[56] Hermann von Helmholtz, "The Origin and Meaning of Geometrical Axioms (II)," *Mind*, 3
(1878):212–13.

the case of the assumption of a "transcendental geometry," he summed up his position as follows:

> The assumption of a knowledge of axioms by transcendental intuition apart from all experience is a) an *unproved* hypothesis, and b) an *unnecessary* hypothesis, since it explains nothing in our actual knowledge of the outer world that cannot equally be explained without its help . . . c) a wholly *irrelevant* hypothesis, since the propositions it includes can be applied to the relations of the objective world only after their objective validity has first been independently proved.[57]

The arguments in this quotation are plainly those of a man primarily concerned with experimental science. From this vantage point, Jevons's or Land's insistence on the existence of a transcendental spatial reality was almost incomprehensible. However, in nineteenth-century England, the nature of geometrical truth was not an issue relevant only to experimental science. There were a variety of other contexts in which the nature of geometrical truth was also germane. The reluctance of English mathematicians to accept Helmholtz's perspective is not best understood as an internal phenomenon of either mathematical or philosophical development. It was bound up in the broader intellectual context within which mathematics was interpreted and pursued.

THE INTELLECTUAL CONTEXT OF GEOMETRICAL DISCUSSION

In the 1860s and 1870s, the cultural context of geometrical development was still the unitary view of truth articulated by men like Herschel and Whewell and integrated by them into the structure of the Cambridge mathematical education. In the middle of the century when the Tripos was reformed, there was dissension among

---

[57] Ibid., p. 225.

England's mathematicians about the specific ways that mathematical truth was to be interpreted and taught. There was remarkably strong agreement, however, about the basic presupposition that mathematics was intimately intertwined with all other subjects in a single tapestry of human knowledge and truth. This conviction remained strong and provided an important context for mathematical work throughout the second half of the century.

The unified view of truth they defended had a significant impact on the way the English viewed non-Euclidean work. Within any such integrated view of knowledge, reluctance to modify any part of the picture would be understandable simply as intellectual inertia, since any change would ripple through all other fields as well. In the case of mathematics in England in the 1860s and 1870s, however, the resistance to change went beyond intellectual lassitude. It was a product of both the particular position mathematics played in the Victorian intellectual scheme and the particular historical circumstances in which the issue was considered.

Within the particular configuration of knowledge the nineteenth-century English had constructed, mathematical truth was central to theology. The transcendental truth mathematics was believed to describe had long stood as the exemplar of the perfect truth to which the human intellect aspired. This was a particularly central issue in theology, where it was often argued that knowledge of the divine partook of the same transcendental necessity as knowledge of mathematics. In this tradition, knowledge of God was defended by being equated with the unquestioned status of geometrical truth.[58]

This theological context made mathematical discussions particularly controversial. Their relevance was further heightened by the period during which non-Euclidean ideas were first widely discussed. Darwin's *On the Origin of Species,* published in 1859, raised a number of serious question for English views of humanly knowable truth. The *Origin* severely threatened the tradition of natural theology, not only because of the bloodthirsty nature Darwin portrayed there, but also because of the kind of truth he implied was

---

[58] A fuller discussion of this context is to be found in Imre Toth, "Gott und Geometrie: Eine viktorianische Kontroverse," *Evolutionstheorie und ihre Evolution,* Dieter Henrich, ed. (Schriftenreihe der Universität Regensburg, band 7, 1982) pp. 141–204.

to be attained through science. Although he apparently had spent twenty years trying to fit his theory into some form of Herschel's and Whewell's framework for legitimate science, Darwin's work did not fit the model of scientific truth they had constructed. His *Origin* contained a highly persuasive argument, but, by traditional standards, Darwin did not seem to have proved the *truth* of his theory.

Many of Darwin's critics used this argument as a way to discredit his work. They dismissed it as perhaps interesting but not adequately demonstrated to be acceptable. Some of Darwin's supporters, on the other hand, turned the argument on its head and attacked the view of scientific truth on which his detractors took their stand. They agreed that Darwin had not proved his theory scientifically in the way such a proof was generally defined. However, they went on to question whether such proofs were ever possible. They challenged the whole notion of scientific truth on which Whewell and Herschel had relied so heavily.[59]

In the 1860s and 1870s, a number of scientists, including, for example, Thomas Henry Huxley and John Tyndall, proposed a new perspective on science and culture. Their new intellectual synthesis was erected on the same basic premise their predecessors had embraced, that the scientific method was an adequate method—in fact, the only legitimate method—for all intellectual inquiry. What set this group of "scientific naturalists" apart from both Herschel and Whewell, and made them threatening, was their strict delimitation of the kinds of truth to which scientific investigation could lead. Their position was very close to that above ascribed to Helmholtz. Specifically, the scientific naturalists claimed it was impossible to arrive at true knowledge of any reality which lay beyond or behind our sense perceptions. They firmly maintained that people could only claim to know the information received through the senses. Such knowledge was always contingent and limited by the field of possible experience. Transcendental realities, the scientific naturalists insisted, were unknown and unknowable in any field.

The range of issues involved in this point of view can be illustrated by looking at alternative interpretations of the notion of "cause"

---

[59] For a fuller development of this interpretation of the response to Darwin, see David L. Hull, *Darwin and His Critics,* (Chicago: University of Chicago Press, 1973).

which were argued at this time in the English periodical literature.[60] The meaning of a causal statement, for example that gravity caused rocks to fall, was the focus of considerable debate. One group, the "empiricists," argued that our knowledge of causal connection did not extend beyond the simple observation that one event was inevitably followed by another. So, for the empiricists, the term gravity was merely a linguistic convenience for summarizing observations about what happened when rocks were dropped. It was not real in the same way that the rock was. The "idealists," on the other hand, maintained that there was an unobservable force or power, a real cause, which tied causally related events together. That such causes, like gravity, were not directly observable did not imply that they were not real. They just could not be observed directly.

The root of the disagreement between the empiricists and the idealists lay in the question of whether man could have any knowledge beyond that of direct experience. The empiricists emphasized that all knowledge came through the senses. Therefore humans could never come to know anything they could not sense. In particular, they could never hope to know a transcendental essence of experience which lay behind the observable facts. The idealists, on the other hand, hung tightly to their conviction that transcendental knowledge was possible. This stand required that they recognize sources of knowledge in addition to the senses. In the idealists' view, indirect reasoning, intuitions, revelation or some such nonexperiential route could give real knowledge.

When conjoined with the Victorian convictions that knowledge was an essentially seamless whole and that scientific knowledge could serve as a norm of truth, the argument between the empiricists and idealists took on very powerful theological implications. In traditional natural theology, God was known to exist because of the traces he had left behind. So, for example, the extraordinarily complex and functional eye pointed directly to its cause: a designing God. The empiricists' strict limitation of knowledge made this kind of inference invalid. One could not assert the reality of the cause by

---

[60] Alvar Ellegaard, *Darwin and the General Reader: The Reception of Darwin's Theory of Evolution in the British Periodical Press, 1859–72*, Gothenburg Studies in English, 8, (Goteborg, 1958).

knowledge of the effects in this case any more than in physics. What is more, since all of knowledge was scientific, this essentially closed the question. In the view of the scientific naturalists, knowledge of God was simply unattainable. Huxley, one of the most prominent spokesmen for this point of view, coined the term "agnostic" to describe this conclusion.

The centrality of these issues to the post-Darwinian English intellectual community is clearly illustrated by the formation of the Metaphysical Society in 1869. This group of prominent scientists, theologians and literary men was formed to engage in free and open debate in an attempt to ease the tensions being generated between science and religion. Its membership was illustrious and represented a broad spectrum of opinion running all of the way from the Catholic theologian W. G. "Ideal" Ward to Huxley, "Darwin's bulldog." The very notion that such a group could constructively discuss and perhaps ultimately agree on issues concerning the nature of knowledge, belief and, ultimately, truth is strong testimony to the depth of the English belief in the unified, ultimately singular nature of those things which are true.[61]

The discussions within this society tended to split along the empirical/ideal lines above illustrated in considerations of the nature of cause. Questions about the nature of mathematical truth were often central to the discussion. This is particularly clear in an interchange which was touched off by a paper, entitled "Necessary Truth," delivered to the society by James Fitzjames Stephen, Leslie Stephen's brother. In this paper Stephen examined the claims that there was such a thing as a necessary truth and found them wanting. Therefore he denied that such truth was real. Stephen's arguments were promptly countered by Ward in his journal, *The Dublin Review*. This prompted Stephen to print his paper in *The Contemporary Review*, and the discussion continued there.

The question of the existence of necessary truth was of paramount importance to Ward. He was in the middle of writing a series of articles, the purpose of which he described as "to establish

---

[61] For a fuller discussion of this society and its activities see Alan Willard Brown, *The Metaphysical Society, Victorian Minds in Crisis, 1869–1880*, (New York: Columbia University Press, 1947).

securely on argumentative ground, against the antitheists of this day, the existence of that Personal and Infinitely Perfect Being, whom Christians designate by the name of 'God.' . . . Now, among the earliest and most essential steps in this argument," Ward continued, "is the thesis, that certain truths are cognizable by mankind as 'necessary.'"[62] Stephen was just the kind of "antitheist" Ward was attacking with his arguments. Stephen's counterattack against the notion of necessary truth could not be ignored.

A central issue their concern with necessary truth led Ward and Stephen to consider was the nature of mathematical truth. Neither man was remarkably proficient mathematically, so, not surprisingly, this part of their discussion is not particularly novel. Ward insisted on the necessary truth of geometrical statements, using as his example, "all trilateral figures are triangular."[63] This statement, he asserted, was true in such a way that even omnipotent God could not change it. Stephen argued essentially that insofar as such statements were necessary they were linguistic tautologies, and insofar as they were true they were learned, like all truths, from experience. This part of the exchange is more useful for illustrating the importance of mathematical truth in theological argument than for providing new angles on that truth itself.

Stephen's and Ward's papers are also useful for showing the role of non-Euclidean geometry in the context of the post-Darwinian discussions of science and religion. Although he was not conversant with their complexities, Stephen *did* know of non-Euclidean developments. He also knew that they had important implications which undercut Ward's position on necessary truth. The source of this conviction was Clifford, who, in 1874, was elected to the Metaphysical Society. Although perhaps he did not understand the technical details from which Clifford was arguing, Stephen, like all of his contemporaries, could find an easily comprehensible interpretation of the meaning of the new geometries in the mathematician's prolific writings.

---

[62] William George Ward, "Necessary Truth [In Answer to Mr. Fitzjames Stephen]," *Contemporary Review*, 25 (1874–75): 527.
[63] Ibid., p. 537.

Clifford fell unambiguously on the empirical side of the arguments over scientific truth in the 1860s and 1870s. Although he is not as well known as Huxley or other scientific naturalists, among his contemporaries he acted as a potent force. He was the youngest of the proponents of the new ideas; he was very iconoclastic and uncontrolled in his attacks on the traditional values and institutions of his compatriots. His ideas were powerful, his presentation of them equally so, and they appeared in many of the most well-known periodicals of the day.

Clifford, the polemicist for scientific naturalism, and Clifford, the mathematician, have, for the most part, been treated separately and scantily by historians, who are either investigating the religious, epistemological and moral issues of the day or are pursuing the mathematics of the period. In many of his writings, however, these two aspects of the man are intimately connected. The new view of geometrical truth, which was involved in the non-Euclidean ideas of the metric geometers, was an integral part of Clifford's arguments for scientific naturalism. In his writings he clearly spelled out the implications of the new geometry in the context of English post-Darwinian discussion.

The position the new geometry played in Clifford's agnostic worldview is evident in a lecture delivered in 1872, "On the Aims and Instruments of Scientific Thought." Here he clearly set forth his characteristically Victorian conviction that human knowledge of all things must come through the application of scientific thought. In his words: "The subject of science is the human universe; that is to say, everything that is, or has been, or may be related to man."[64]

Clifford quickly went on to establish the limits of this claim. Scientific knowledge, he emphasized, can never extend beyond the limits of what we have directly experienced. Consequently, it can never allow us to attain an exact knowledge of an absolute or transcendental realm which lies behind the world as we know it. Clifford clarified and emphasized this point by drawing a careful distinction between contingent knowledge, which he labeled as "practically

---

[64] W[illiam] K[ingdon] Clifford, "On the Aims and Instruments of Scientific Thought," *Macmillan's Magazine*, 26 (1872): 499.

exact," and absolute knowledge, which he labeled as "theoretically exact." Practical exactness, he stated, could be attained by physical measurement. It was the exactness affirmed by an honest grocer who claimed to have measured exactly a pound of sugar; it was a measurement accurate to the limits of his scales. Theoretical exactness was of a totally different kind. Clifford described it as follows:

> Suppose the mass of the standard pound to be represented by a length, say a foot, measured on a certain line; so that half a pound would be represented by six inches, and so on. And let the difference between the mass of the sugar and that of the standard pound be drawn upon the same line to the same scale. Then, if that difference were magnified an infinite number of times, it would still be invisible. This is the theoretical meaning of exactness; the practical meaning is only very close approximation.[65]

Theoretical exactness was the kind of exactness which was claimed in mathematics, specifically geometry. When a mathematician made a statement about congruence, he was not confining his claims to any limits of error. He was dealing with theoretical exactness.

Theoretical exactness posed a problem for the strict empiricism Clifford and the other scientific naturalists espoused, however. This is because it implied a knowledge which transcended finite experience. "The knowledge then of an exact law in the theoretical sense would be equivalent to an infinite observation," Clifford wrote. He went on to deny that it was a kind of knowledge which humans knew. "I do not say that such knowledge is impossible to man; but I do say that it would be absolutely different in kind from any knowledge that we possess at present."[66]

Clifford acknowledged that denying theoretical exactness was a surprising position for a mathematician to take, since mathematics was the paradigm field for theoretically exact statements. "I shall be

---

[65] Ibid., pp. 503–4.

[66] Ibid., p. 504. That Clifford left open the possibility of theoretically exact knowledge is merely an acknowledgement of agnosticism, here as everywhere. His subsequent writings clearly indicate that he did not take this possibility seriously. In fact he seems to have made almost a vocation of being an intellectual watchdog, attacking anyone with the audacity to draw transcendental conclusions from scientific work.

told, no doubt, that we do possess a great deal of knowledge of this kind, in the form of geometry and mechanics; and that it is just the example of these sciences that has led men to look for exactness in other quarters." He went on, however, to emphasize that the development of non-Euclidean geometry had demonstrated just how mistaken the view of mathematical exactness was.

> It happens that about the beginning of the present century the foundations of geometry were criticised independently by two mathematicians, Lobatchewsky and the immortal Gauss; whose results have been extended and generalized more recently by Riemann and Helmholtz. And the conclusion to which these investigations lead is that, although the assumptions which were very properly made by the ancient geometers are practically exact—that is to say, more exact than experiment can be—for such finite things as we have to deal with, and such portions of spaces as we can reach; yet the truth of them for very much larger things, or very much smaller things, or parts of space which are at present beyond our reach, is a matter to be decided by experiment, when its powers are considerably increased.[67]

For Clifford, non-Euclidean geometry had demonstrated that mathematical truths, like any other, were only practically exact.

By restricting mathematics to practical exactness, Clifford was severely limiting the amount human beings could hope to know. He made this point explicitly in his lecture "The Postulates of the Science of Space." Here Clifford pointed to the Copernican revolution as forcing a radical limitation to man's perception of his knowledge of the universe. Before Copernicus, the universe was conceived to be a bounded and knowable entity, existing in a boundless eternity which itself was completely described and understood. Copernicus's formulation of cosmology shattered this conception of a closed, knowable world. In its stead he introduced an infinite universe about which man knew very little. Rather than claiming to know all about what there was, man could claim to know only about the little part of the universe in which he dwelled.

---

[67] Ibid., p. 504.

Even after Copernicus had effected this fantastic change of outlook, however, there remained one area where man still knew everything about the universe—in the laws of space and motion. Clifford
explained the position as follows:

> For the laws of space and motion . . . implied an infinite space and
> an infinite duration, about whose properties as space and time every-
> thing was accurately known. The very constitution of those parts of
> it which are at an infinite distance from us, 'geometry upon the plane
> at infinity,' is just as well known, if the Euclidean assumptions are
> true, as the geometry of any portion of this room. In this infinite and
> thoroughly well-known space the Universe is situated during at least
> some portion of an infinite and thoroughly well-known time. So that
> here we have real knowledge of something at least that concerns the
> Cosmos; something that is true throughout the Immensities and the
> Eternities.[68]

Non-Euclidean geometries had destroyed this small piece of universal and absolute knowledge, which had remained in the Newtonian world. Because there were alternative geometrical possibilities,
humans could not claim to know anything about places or magnitudes they had not directly experienced. In the construction of non-
Euclidean geometries Clifford found the destruction of human
hopes for the kind of universal knowledge in *any* area which had
been initially claimed in geometry. He wrote,

> So, you see, there is a real parallel between the work of Copernicus
> and his successors on the one hand, and the work of Lobatchewsky
> and his successors on the other. In both of these the knowledge of
> Immensity and Eternity is replaced by knowledge of Here and Now.
> And in virtue of these two revolutions, the idea of the Universe, the
> Macrocosm, the All, as subject of human knowledge, and therefore
> of human interest, has fallen to pieces.[69]

Clifford rejoiced in the freedom this geometrical variety and consequent uncertainty afforded him. He exuberantly speculated about

---

[68] Clifford, "Postulates," (1875):362.
[69] Ibid., p. 363.

the possible relations between spatial curvature and mass.[70] Contemplating the possibility of a finite, curved universe, he told his audience at the Royal Institution: "In fact, I do not mind confessing that I personally have often found relief from the dreary infinities of homaloidal space in the consoling hope that, after all, this other [bounded space] may be the true state of things."[71]

He also accepted the agnosticism, which for him shaded into atheism, which his views implied. He firmly defended it throughout his fight against tuberculosis which killed him in his thirty-first year. The epitaph he composed for himself, now hidden in Highgate's weeds, unequivocally affirms his refusal to speculate past accessible experience.

> I was not, and was born.
> I loved and did a little work.
> I am not and grieve not.[72]

Clifford's clearly stated and highly radical presentations of the implications of non-Euclidean work clarify the situation facing the English with the introduction of the new mathematics from the continent. Riemann's and Helmholtz's analytical constructions seemed to be directed against the very heart of a view of mathematical truth which was central to an entire intellectual and institutional tradition of mathematical work. For the most part, England's mathematicians did not share in Clifford's excitement at destroying the previous interpretation of the nature of mathematical truth. To quote again from Roberts:

> I cannot regard it as otherwise than unfortunate that the unqualified transference of language, of which I speak, should take place, not as a modest matter of convenience, nor as a concession to the mental limitations which vex the dreams of those who seek a perfect language, but with pomp and blow of trumpet, as a signal victory of the empirical school.[73]

---

[70] Clifford, "Space-Theory," (1876):157–58.
[71] Clifford, "Postulates," (1875):376.
[72] Highgate Cemetery, grave no. 23181.
[73] Roberts, "Remarks," (1882–83):10.

The solution to this problem was not far to seek, however. By 1883, when he addressed the BAAS, Cayley was able to dismiss the whole question as unimportant. The basis of Cayley's position was yet another approach to non-Euclidean geometry, different from that of either the synthetic or the metric geometers. This third point of view, the projective point of view, fit the various needs of England's mathematicians singularly well. It both conformed to major aspects of the empirical approach to knowledge and left the door open to a view of geometry as necessary truth. In addition, it allowed all non-Euclidean geometries to be interpreted with an integrated framework. The English assimilation of the new and vital subject of projective geometry will be the subject of the next chapter.

"Hand with Reflecting Globe." (© *1988 M. C. Escher c/o Cordon Art—Baarn—Holland.*)

CHAPTER 3

Projective Geometry and Mathematical Science

As first presented by the writings of Helmholtz, Riemann and Clifford, non-Euclidean geometry was a radical mathematical development which threatened to drastically distort the Victorian intellectual tapestry in which geometrical study was so integrally woven. Geometry's place in the mid-century picture of knowledge was bound up with its special truth status. Geometrical theorems were the epitome of necessary truths; truths conceived so clearly as to be indisputable. This status rested on at least two closely related assumptions. The first was the presumed uniqueness of geometry; there was only one way to geometrize space. The other way of making the same point was essentially cognitive; once a geometrical theorem was clearly understood, its truth was so overwhelming that alternatives to it could not even be conceived.

The non-Euclidean developments introduced in the 1860s and 1870s undermined geometry's claims to necessary truth from both the factual and the cognitive points of view. The assumption that there was only one way to geometrize space was undermined by the synthetic geometers working during the first part of the century. Lobachevskii, Bólyai and others claimed to have destroyed the uniqueness of Euclid's geometry simply by producing alternatives. The later metric geometers addressed the cognitive assumption, challenging the whole notion of conceptual clarity which had supported geometrical truth. These synthetic and metric arguments embroiled the English both in discussions about what legitimately constituted a "geometry" and in discussions about what was actually involved in "conceiving." Mathematical skill was not directly relevant to these issues, and yet they were central to the Victorian view of knowledge in which geometry was both institutionally and intellectually embedded. Whatever their opinions on the specific issues raised, non-Euclidean developments generated significant pressure for those in England's mathematical community.

The period of intense pressure was rather short-lived, however. Using Sommerville's *Bibliography of non-Euclidean Geometry* as a guide, general interest in non-Euclidean issues peaked in about 1878 and declined after that. An explanation for this decline can be found in the growing strength of a third approach to the new geometries, an approach that interpreted these geomtries in a significantly different way from either the synthetic or the metric approaches.

This chapter gives a four-part account of the development of projective geometry during the first part of the nineteenth century. The first section outlines the origins of the study as it developed first in France and later in England up to the time of Darwin. The second section examines the ideological and epistemological framework within which the study of projective geometry was embedded in post-Darwinian England. The third section traces the assimilation of significant non-Euclidean developments into both the mathematical and ideological structure of projective geometry. This interpretative process defused much of the controversy raised by earlier non-Euclidean developments but did not wholly resolve more basic tensions raised within the conceptual interpretation of geometrical truth. These tensions, and the process of their resolution in late-Victorian England, are the subject of the final section.

THE EARLY DEVELOPMENT OF PROJECTIVE GEOMETRY

The focused study of projective geometry began in France early in the nineteenth century. Scattered theorems, now recognized as part of the subject, had been demonstrated before—notably by Pascal and Desargues—but sustained interest in projective geometry was first stimulated by the work of Gaspard Monge, after the revolution.

Monge's work originated as a practical technique for reducing complex three-dimensional problems to more manageable two-dimensional ones. His book, *Géométrie descriptive*, contains a method by which essential geometrical features of three-dimensional figures can be transformed and incorporated in two-dimensional drawings. The method Monge used to describe solids on the plane (the method of double orthographic projection) created a virtual revolution in military engineering design and is still a central part of mechanical drawing.[1]

In addition to its practical value Monge's method of projection suggested a whole new way to look at geometrical figures. In the hands of his many brilliant pupils, like Jean-Victor Poncelet, Charles Dupin and Michel Chasles, Monge's descriptive methods were developed into what is now recognized as "projective geometry": the systematic study of those geometrical relations which remain invariant under carefully prescribed processes of projection and section. To use a modern image, this approach is essentially the study of the families of shapes produced by a projected image as the screen is moved nearer or farther or turned to different angles. As developed by this group of Frenchmen, the study of projective geometry was linked to a particular epistemological stance. These mathematicians were responding to serious difficulties they saw lurking in the foundations of the burgeoning study of analysis.

Analysis presented particularly pressing philosophical problems in France early in the nineteenth century. Within the epistemology of Locke and his French interpreter, Condillac, knowledge was introduced through the senses. Raw sensations were received into the mind where they were analyzed and organized in a variety of abstractive ways. These mental processes were highly effective in generating new ideas. Ultimately, though, the validity of any idea rested on concrete experience. In this view, abstractions might be suggestive, but they were far from reality.

This sensationalist epistemology was similar to mid-nineteenth-century English interpretations of knowledge because it emphasized

---

[1] Gaspard Monge, *Géométrie descriptive, Leçons données aux écoles normales, l'an 3 de la République* (Paris: Baudouin, an viii [1798]).

that mathematical truth rested on an exact correspondence between mathematical ideas and something these ideas described. The major difference was that in eighteenth-century France, the thing described tended to be a physical object or a sensation, whereas in mid-century England it was more often a concept. In both cases, however, geometry as opposed to analysis was the quintessential mathematical subject.

This was because the objects of geometrical reasoning were clear and distinct, its processes easily identifiable in the outside world. The foundations of analysis, on the other hand, were very doubtful. The analytic operating procedure often seemed to involve mere symbol juggling rather than clear thinking; furthermore, it was often highly unclear what the symbols meant. Thus, in post-revolutionary France, geometry was as foundationally superior to analysis as it was in mid-century England.

However, serious difficulties were encountered with this attitude in France early in the nineteenth century. In the eighteenth century, analysis had been developed into a tremendously powerful tool for the development and expression of Newtonian cosmology. Its success in this area raised questions about the validity of treating it as a weaker study than geometry. This pressure to reinterpret the value of analysis was strengthened by the fact that geometry was apparently moribund; it had been essentially unchanged since the time of Euclid. This stagnant state was seen to be related to geometry's specificity. In Euclid's form, geometry lacked all of the suggestive generality which made algebra so powerful, because its theorems and proofs had to be tied to the particular case. Thus, the same feature which made geometry rigorous seemed to make it boring as well.

According to the advocates of projective geometry, Monge's descriptive methods addressed these issues because they made geometrical theorems more general without jeopardizing its concrete rigor. As first practiced, the source of generality in projective geometry was the "principle of continuity," explicitly stated by Poncelet in 1822. This principle asserted that theorems proved for one figure were equally true for correlative figures formed from the original by a continuous transformation. For example, the principle of continuity allowed one to conflate Euclid's proposition (Book III, prop.

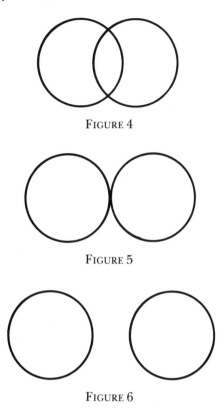

FIGURE 4

FIGURE 5

FIGURE 6

10) "A circle does not cut a circle at more points than two"[2] with his proposition (Book III, prop. 13) "A circle does not touch a circle at more points than one."[3] From the projective standpoint, these propositions were identical because figure 4 can be transformed into figure 5 by a continuous transformation. In this transformation, the intersecting circles slide apart to become tangent, and the two points of intersection become coincident. Accepting the principle of continuity meant that proposition 10 would cover both cases.

As this example suggests, the projective method involved establishing connections among propositions previously considered sep-

---

[2] *The Thirteen Books of Euclid's Elements,* Introduction and Commentary by Thomas L. Heath, 2nd ed. rev., 3 vols. (New York: Dover Publications, 1956), 2, p. 23.
[3] Ibid., p. 32.

arate. These connections created generality, which was perceived as the real advantage of the new geometrical study. It meant that geometry could remain closely tied to physical experience while incorporating algebra's general powers. It also promised exciting new discoveries within the confines of a physicalist interpretation of foundations. As the 1820s unfolded, much of this promise was fulfilled. Projective geometers generated a host of new results such as the theory of poles and polars, of duality, of homographic projections, of anharmonic ratios and more.

Projective geometry with its concern for concrete rigor was not the only early nineteenth-century French view of the nature of mathematical truth. A group of analysts led by Augustin Cauchy maintained that abstract consistency, rather than concrete interpretability, was the foundation of mathematical truth and rigor. This group flatly rejected the mathematical validity of the principle of continuity as well as the program to generalize geometry advocated by the projective geometers. Such generalizations had proven useful as heuristic devices, they admitted, but their vague enunciation disqualified them from any place among the strict canons of mathematical proof.

Particularly disturbing for the new breed of analysts was the quasi-inductive generalizing which the principle of continuity introduced into mathematical argument. The principle essentially legitimated moving from specific examples to general cases in mathematics as in the natural sciences. Such a tie was welcomed in the physicalist tradition. It was not attractive among the analysts, however, who emphasized the distinction between mathematical and scientific truth. Reporting on a paper in which Poncelet relied on this principle, Cauchy stated firmly, ". . . nous avons reconnu que ce principe n'était, à proprement parler, qu'une fort induction qui ne pouvait être indistinctement appliqué à toutes sortes de questions de Géométrie, ni même en Analyse."[4]

The legitimacy of the quasi-inductive reasoning involved in the principle of continuity was a crucial issue on which hung the fate of the post-revolutionary interest in projective geometry. The principle was absolutely central to the pursuit of the subject. It was not

---

[4] *Proces-Verbaux de l'Academie des Sciences*, 7 (1820–23):46.

merely useful for countering claims of analytic superiority, which rested on the greater generality of analysis. Of greater import was the fact that accepting this principle led to the rise of a radically different vision of space than that of Euclid. In particular it suggested the existence of geometrical entities which had no place in Euclid's space: the "imaginary points" and "points at infinity." The way these new points were introduced into geometrical discussion illustrates the inductive quality of the principle of continuity against which Cauchy so vehemently objected.

The use of the principle of continuity to conflate Euclid's propositions 10 and 13 seems a relatively innocuous shortcut in geometrical argument. The principle was inherently much more radical, however. This can be illustrated by simply extending the argument. When the transformation which led from figure 4 to figure 5 is carried one step further, to figure 6, the principle of continuity continues to emphasize their essential connection. Subscribing to the principle of continuity entails accepting that the points of intersection which are distinct in figure 4 are coincident in figure 5 and "imaginary" in figure 6.

Figure 6 is obviously the most problematic generalization from the Euclidean point of view because here the points of intersection cannot be found in the real plane. In classical Euclidean geometry, the circles would have no points of intersection. From the projective standpoint they would. This is because the power of the projective theory rested on emphasizing the similarities among figures previously seen as distinct. Viewed projectively, figure 6 does not differ significantly from figures 4 and 5, since it can be generated from them. Even though the points of intersection in figure 6 cannot be found in the real plane, accepting their existence allows a comprehensive treatment of real geometrical relations.

The role of the imaginary points of intersection of the circles in figure 6 is perhaps better illustrated in figures 7–9, which include not only the circles but also their "radical axes." The radical axis can be defined as the line through the points of intersection of the circles. This line has a number of interesting properties. For example, if from any point on the radical axis, one were to draw lines tangent to the two circles, the distance along these lines to their tangent points would be equal (figure 10).

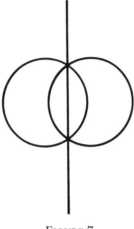

FIGURE 7

In figure 8, the projective method allows the radical axis again to be defined as that line passing through the points of intersection, which are coincident in this case. In figure 9, although the points of intersection are no longer in evidence, one can still find a radical axis with the same properties as those found for figures 7 and 8. The projective geometers' attempt to focus on what they saw as the

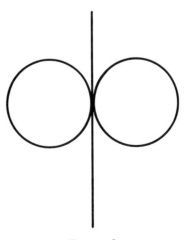

FIGURE 8

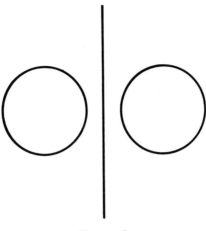

FIGURE 9

deeper relations lying behind the accidents of the individual case led them to see this radical axis as a real line through the imaginary points where the two circles intersect.

The points at infinity, which are somewhat more intuitively satisfying, also fell out of the projective geometers' desire to look beyond the individual to the general. This can be seen in figure 11, where line *A* is transformed by a continuous counterclockwise

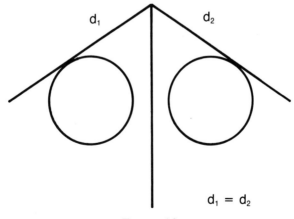

FIGURE 10

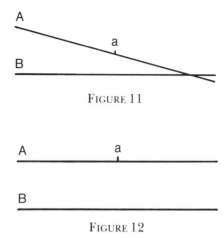

FIGURE 11

FIGURE 12

motion around the fixed point *a*. As this occurs, the point of inter-section of *A* with *B* travels out to the right until the two are parallel. In this state, the principle of continuity asserts that what holds true for figure 11, "two lines intersect in a point," must also hold for figure 12. For this reason, projective geometers recognized a point at infinity which was the point of intersection of two parallel lines. Thus, in projective geometry, parallel lines came to be defined not as lines which never meet, but rather as lines intersecting in a point at infinity.

Imaginary points and points at infinity are not part of the classical concept of space which Euclid analyzed so comprehensively in his *Elements*. However, reasoning based on them introduced a whole new order of generality into geometry. This was the goal of the early French geometers who were striving to make geometry as general, and hence as powerful, as analysis.

Arguments between geometers and analysts raged on the continent throughout the 1820s. In France, analysts ultimately won the day. By the 1840s, the once vital study of projective geometry was only pursued by the isolated Michel Chasles.[5] In mid-century

---

[5] This interpretation of the development of projective geometry in France is more fully pre-sented in Lorraine J. Daston, "The Physicalist Tradition in Nineteenth Century French Geometry," *Studies in the History and Philosophy of Science*, 17 (1986):269–95.

England, however, its fortunes were different. The difference can be accounted for partly by the fact that whereas in France there was support for a self-regulating mathematical community, in England the conflict between the relative value of analytic and geometric concepts of rigor took place primarily within the context of liberal education. Here, in discussions culminating in the 1848 Tripos reform, geometry, with its conceptual foundations, ultimately triumphed as the sturdiest foundation for mathematical truth and the exemplar of rigorous reasoning.

In the English educational context, the geometry emphasized was the elementary geometry of Euclid. There was no mention of projective geometry on the elementary portion of the examination. Yet the triumph of a descriptive over a formal or definitional view of mathematical rigor was not merely educational but philosophical and mathematical as well. The reaffirmation of the value of concrete, "scientific" mathematics which was evident in curricular reform was also reflected in English mathematical work. Mid-century English mathematicians continued their interest in analytic manipulations, notably in the algebra of invariants. But, in principle, all such work required an interpretation to ground it in the real world. This interpretation was often provided by projective geometry. The general theorems generated by the French projective method were ideally suited for the interpretation of general algebraic equations; after all, projective geometry had been first developed with the express goal of concretely grounding such abstract researches. After the middle of the century, in the mathematical journals, projective geometry often served as the interpretation for more sophisticated analytic results.

One of the foremost representatives of this way of thinking was Arthur Cayley, England's most prominent Victorian mathematician. In his "Sixth Memoir upon Quantics," published in 1859, Cayley explicitly considered the relationship between projective geometry and the algebra of invariants. The "Sixth Memoir" was written as one in a series of articles focused on "quantics," homogeneous algebraic polynomials in two or more variables. Of the ten articles, the sixth was the only one addressing geometrical issues. In it, Cayley developed a projective interpretation for the algebraic results he was generating.

For Cayley, the stumbling block for interpreting analytical results projectively was the notion of distance. In Euclidean interpretations of analytic expressions, numbers and variables were routinely translated into coordinates whose value was determined by the distance relations among points. However, the transformations of projection and section by which projective geometry was defined obliterated traditional measures of distance. A line segment, the traditional measure of distance, can be lengthened or shortened at will through projection. Not only can the absolute value of distances be transformed, the relative distances between collinear points can also be distorted. In figure 13, for example, the relationship between AB and BC is different from that of its projections A_oB_o and B_oC_o. Thus, the challenge to the interpretive role of projective geometry Cayley faced in his "Sixth Memoir" was to find a function which exhibited the basic properties of the distance function and yet was invariant through the transformations of projection and section.

Cayley found the solution to this challenge in the cross ratio, a relationship among points which projective transformations leave unaltered. This ratio is formed as follows. Let A, B, C, and D be four collinear points. Consider the ratio in which B divides the line AD, AB/BD, and the ratio in which C divides AD, AC/CD. The quotient of these ratios, $AB/BD:AC/CD$, is invariant under projective transformation. This quotient is called the cross ratio. Thus in figure 13,

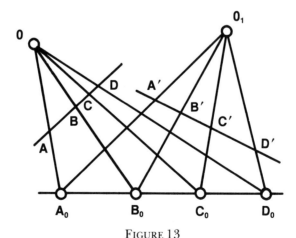

FIGURE 13

projecting line AD onto line A_oD_o might well radically change the distances among the corresponding points. However, no matter what combinations of projection and section are used, the cross ratio in which B and C divided AD would be the same as that in which B_o and C_o divided A_oD_o:

$$BA/CA\!:\!BD/CD \;=\; B_oA_o/C_oA_o\!:\!B_oD_o/C_oD_o.$$

Faced with the lack of an invariant distance relation in projective geometry, Cayley proceeded to construct an analogue on the basis of the invariant cross ratio. In a projective space of two dimensions, he hypothesized the existence of a conic which he termed the Absolute. Any line in the plane would intersect the Absolute in two points, y and z. The distance between any two points P and Q on the line, was defined as a function of the cross ratio of P, Q, and the Absolute points x and y. Cayley verified that the function he thus defined fit the same system of algebraic relations as the conventional distance function. For example, he verified that given three points on a line, P, P′, P″, his function implied that:

$$\text{dist (PP′)} + \text{dist (P′P″)} = \text{dist (PP″).}$$

Since his newly defined function exhibited the same algebraic properties as distance, Cayley felt justified in regarding it as the distance function in projective space. With the identification of such a distance function he felt he had created a metric geometry within essentially nonmetric projective space. Cayley's success in generat-

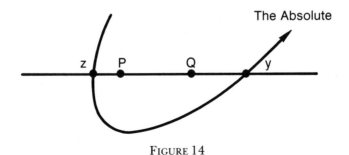

FIGURE 14

ing metrical geometry from the amorphousness of projective space led him to the conclusion that projective geometry (which he called "descriptive" geometry) was fundamentally prior to classical Euclidean geometry. He concluded his memoir on this note:

> the theory in effect is, that the metrical properties of a figure are not the properties of the figure considered *per se* apart from everything else, but its properties when considered in connexion with another figure, viz. the conic termed the Absolute. The original figure might comprise a conic; for instance, we might consider the properties of the figure formed by two or more conics, and we are then in the region of pure descriptive geometry: we pass out of it into metrical geometry by fixing upon a conic of the figure as a standard of reference and calling it the Absolute. Metrical geometry is thus a part of descriptive geometry, and descriptive geometry is *all* geometry, and reciprocally.[6]

Cayley saw himself as having moved the understanding of geometrical space a giant step forward with his definition of distance. Euclidean geometry, firmly rooted in the notion of distance, had for centuries been considered the fundamental description of space. With his projective theory of distance Cayley thought he had revealed the more basic, hitherto hidden spatial structure lying behind this metrical construction. From his point of view, classically conceived geometrical space was essentially a refinement of a more fundamental projective space which had been revealed through the study of projective geometry.

The remainder of this chapter will focus on the English assimilation of projective geometry in the period following Cayley's "Sixth Memoir." This paper of 1859 is a convenient point from which to date a surge of interest in projective geometry in the second part of the century. Part of this convenience lies in its temporal coincidence with Darwin's *On the Origin of Species,* a coincidence suggesting a connection between the two works. In terms of their reception, this connection is real. The fortunes of the geometry which Cayley suggested was *all* geometry—during this period indiscriminately

---

[6] Arthur Cayley, "A Sixth Memoir upon Quantics," *Philosophical Transactions of the Royal Society,* 149 (1859):61–90, reprinted in *The Collected Mathematical Papers of Arthur Cayley,* 11 vols., (Cambridge: University Press, 1889–97), 2, p. 92.

referred to as "descriptive," "modern," "projective" or even "the newer" geometry—were linked to important themes in post-Darwinian English scientific life.

MODERN GEOMETRY AND ENGLISH CULTURE

Perhaps the most common of the many terms used to designate projective geometry in late nineteenth-century England was "modern geometry." This term accurately captures the spirit of excitement which surrounded the study in the 1860s and 1870s. These decades witnessed the literal demise of the scions of the Analytical Society and the rise to power of a new generation of English mathematicians.

The outline of the new community can be charted from the diary of T. Archer Hirst, a continentally trained mathematician who followed Augustus De Morgan as professor of mathematics at University College in London. Hirst's world contained such mathematical greats as Arthur Cayley, J. J. Sylvester and H. J. S. Smith as somewhat older members. The solid establishment these men represented was balanced by a rising younger set which included Hirst himself, the young Dane Olaus Henrici, William Spottiswoode and the flamboyant "lion of the season"[7] William Clifford.

Although their primary interests were mathematical, these men were active members of the larger intellectual community of post-Darwinian England. Hirst, for example, was as involved with the BAAS and the Royal Society as he was with the infant London Mathematical Society. His research interests were clearly mathematical, but they did not isolate him from the heady world of the Metaphysical Society or the other activities of the English scientific community.

In the last chapter, the vital world of post-Darwinian science was treated primarily from an intellectual perspective, which emphasized the controversy between idealist and empirical philosophies. One way to understand the urgency of the arguments between adherents to these different points of view is illustrated by the theological problems generated by a strictly empirical stance. However,

---

[7] William H. Brock and Roy M. MacLeod, *Natural Knowledge in Social Context: The Journals of Thomas Archer Hirst, S. R. S.*, (London: Mansell, 1980), p. 1828.

the scientific ferment in late-Victorian England is not adequately understood by merely focusing on abstruse intellectual discussions about God and necessary truth. Much of what concerned men like Huxley and Tyndall, or their more mathematical friends like Hirst and Clifford, was that science in England needed to be more adequately supported. This new generation of scientists rebelled against the financial implications of the earlier gentlemanly ideal and argued for the importance of substantial support for scientific research.[8]

One of the defining characteristics of this group's vision was a keen interest in research and a strong desire to import some variant of the German research ideal into England. This meant emphasizing scientific change and development as positive ideals. This position was at odds with the tradition of humanistic science espoused by Whewell and his followers. In his writings about the liberal education, Whewell had emphasized the importance of at first introducing students only to the most perfect examples of human achievement. Their characters would be most effectively formed by exposure to the best fruits of human endeavor. Among other things, this approach led Whewell to distinguish between progressive and permanent studies. Progressive studies were, as their name suggests, those in which ideas were still changing and being developed. Permanent studies, on the other hand, were those which were complete and unchanging, whose value had been consistently attested to by generations. Whewell had insisted that permanent studies were the proper center of a liberal education. Because they were trying to develop the highest human qualities in the students by acquainting them with those qualities in others, "nothing less than the most thoroughly luminous and transparent [understanding] will suffice."[9] Students ought not be exposed to the kind of confusion often

---

[8] For a more detailed analysis of the social issues these men were battling see Roy MacLeod, "Resources of Science in Victorian England: The Endowment of Science Movement, 1868–1900," *Science and Society 1600–1900,* Peter Mathias ed., (Cambridge: University Press, 1972), pp. 111–66. An analysis of the connection between these social questions and some of the intellectual questions in which they were embroiled is to be found in Frank M. Turner, "The Victorian Conflict between Science and Religion: A Professional Dimension," *Isis,* 69 (1978):356–76, or in Frank M. Turner, *Between Science and Religion,* (New Haven: Yale University Press, 1974), pp. 1–37.
[9] William Whewell, *Philosophy of the Inductive Sciences,* 2nd ed. rev., 2 vols., (London: 1847; reprint ed., London: Frank Cass & Co. Ltd., 1967), 2, p. 366.

involved in progressive studies until the more advanced stages of an education.

The younger, primarily London-based group of post-Darwinian scientists who were lobbying for the active support of scientific research opposed the static scientific ideal epitomized by the emphasis on permanent studies in the Cambridge education. They countered claims for the pre-eminent value of permanent truth with an emphasis on the value of change and the importance of new developments for scientific progress. They argued that the distinguishing value of science lay as much in its methods as in the kinds of truth it generated. This point of view was clearly expressed in many of Huxley's writings promoting English science.

Thus, for example, in "An After-dinner Speech" delivered to the Liverpool Philomathic Society, Huxley proclaimed the unique value of a scientific education for practical life and as essential background to learned professions. In addition, he extolled the virtues of basic scientific training as preparation for *any* intelligent adult life. As he put it:

> I do not mean that every schoolboy should be taught everything in science. . . . What I mean is that no boy nor girl should leave school without possessing a grasp of the general character of science, and without having been disciplined . . . in the methods of all sciences; so that, when turned into the world to make their own way, they shall be prepared to face scientific discussions and scientific problems . . . by being familiar with the general current of scientific thought, and being able to apply the methods of science in the proper way.[10]

The distinguishing feature of this scientific method of dealing with problems was its immediate observational nature, the "bringing of the mind directly into contact with fact, and practising the intellect in the completest form of induction; that is to say, in drawing conclusions from particular facts made known by immediate observation of nature."[11] The ability to observe the relevant particulars and to size up a situation according to its particular merits was

---

[10] Thomas H. Huxley, "Scientific Education: Notes of an After-dinner Speech," *Macmillan's Magazine,* 20 (1869):181.

[11] Ibid., p. 182.

the key skill which Huxley hoped to develop through scientific education, a skill which he felt would be relevaent to all kinds of problems the student would meet in later life.

Huxley's vision was an exciting one which appealed to many young scientific thinkers. It was more problematic for mathematical enthusiasts, however, because it threatened to degrade their subject. As Huxley presented it, mathematics, "that [study] which knows nothing of observation, nothing of experiment, nothing of induction, nothing of causation,"[12] was not one of the sciences but rather handmaiden to them all. He glided over its claims to transcendental truth by emphasizing its subjective over its objective character. Its axioms, he wrote, were "called self-evident." But essentially, in the study of pure mathematics as in the study of language, "authority and tradition furnish the data, and the mental operations of the scholar are deductive."[13] Thus construed, mathematics was empty, esoteric and of no general intellectual value.

The mathematical response to Huxley's allegations that mathematics was not suited to education was virtually immediate. In August of 1869, James J. Sylvester, one of Britain's leading mathematicians, entitled his Presidential Address to the Mathematics and Physics Section of the British Association for the Advancement of Science, "A Plea for the Mathematician." In this work he vehemently attacked Huxley's characterization of mathematics, noting that perhaps Huxley would have couched his views in "more guarded terms . . . had his speech been made before dinner instead of after."[14]

Sylvester was somewhat older than Huxley and had received some of his education at Cambridge. Because he was Jewish, however, he was never allowed to fully participate in the world of the Cambridge educational vision. Perhaps as a result, he did not defend the absolute fixed truth of mathematics. Instead, he mounted his defense of

---

[12] Thomas H. Huxley, "The Scientific Aspects of Positivism," *Fortnightly Review*, n.s. 30 (1869):667.

[13] Huxley, "Notes," (1869):182.

[14] James J. Sylvester, "Presidential Address: Mathematics and Physics Section," *Report of the Thirty-ninth Meeting of the BAAS Held at Exeter in August 1869*, (London: John Murray, 1870):1–9; reprinted as "A Plea for the Mathematician" in *Nature*, 1 (1869–70):237.

mathematics within the same progressive camp which supported Huxley's attack. Rather than defending the value of Whewell's permanent studies, he argued that much of mathematics was highly progressive.

In his "Plea" Sylvester painted a highly naturalistic view of mathematics. He countered Huxley's allegation that the study of pure mathematics was essentially deductive by saying:

> mathematical analysis is constantly invoking the aid of new principles, new ideas, and new methods . . . springing direct from the inherent powers and activity of the human mind, . . . it is unceasingly calling forth the faculties of observation and comparison, . . . one of its principal weapons is induction, . . . it has frequent recourse to experimental trial and verification, and . . . affords a boundless scope for the exercise of the highest efforts of imagination and invention.[15]

He continued, vividly describing example after example where mathematicians had relied on "the faculty of observation" to come to their results, as in the "accidental observation by Eisenstein . . . of a single invariant . . . which he met with in the course of certain researches just as accidentally and unexpectedly as M. Du Chaillu might meet a Gorilla in the country of the Fantees, or any one of us in London a White Polar Bear escaped from the Zoological Gardens."[16]

In thus defending his subject, Sylvester explicitly decried the finished, static emphasis of traditional mathematical education. He welcomed Huxley's suggestion that the spirit of open-ended, inductive sciences be made central in its stead. Mathematical education, he emphasized, could easily be adjusted to that vision. "I should rejoice," he exclaimed,

> to see mathematics taught with that life and animation which the presence and example of her young and buoyant sister [inductive science] could not fail to impart, short roads preferred to long ones, . . . projection, correlation, and motion accepted as aids to geometry;

[15] Ibid.
[16] Ibid., p. 238.

the mind of the student quickened and elevated and his faith awakened by early initiation into the ruling ideas of polarity, continuity, infinity, and familiarisation with the doctrine of the imaginary and inconceivable.[17]

Sylvester's invocation of the projective ideas of correlation, polarity, continuity, infinity and the imaginary in his defense of geometry reflects a strong trend among his contemporaries. There was a real surge of interest in projective geometry in England in the 1860s and beyond. Hirst's journal consistently returns to the theme of projective geometry. There he laid out in vivid detail the political machinations which led to ultimate success in awarding the 1865 Copley medal to the last remaining French projective geometer, the aging Michel Chasles. Outside of the arena of scientific politics, the sense of excitement Hirst felt surrounded the new geometrical study is often evident in his entries, such as the following of Sunday, February 25, 1866.

> I have had several letters during the week from Cayley on Geometrical Transformation. I wish I were at liberty to do my part in the important investigations that are now ripe . . . Sylvester again is actively thinking and producing, and Chasles has just published a most important extension of his method.[18]

Hirst's privately expressed excitement about projective geometry was echoed in many public statements by England's mathematicians. Thus, for example, in 1865 Spottiswoode's attempt to provide the BAAS with "a short summary of subjects which have engaged the attention of philosophers during the past year"[19] included projective geometry as one of three major areas of mathematical interest. (The other two were the analysis of invariants and the theory of numbers.) More than a decade later in a speech to the London Mathematical Society entitled "On the Present State and Prospects of some Branches of Pure Mathematics," H. J. S. Smith pointed to

---

[17] Ibid., p. 261.

[18] Brook and MacLeod, "Hirst," (1980), p. 1775.

[19] William Spottiswoode, "Address to Mathematics and Physics Section," *Report of the Thirty-fifth Meeting of the BAAS held at Birmingham in August 1865*, (London: John Murray, 1866), p. 1.

an overriding interest among his mathematical colleagues in "that combination of the newer algebra [the algebra of invariants] with the direct contemplation of space which constitutes the modern [projective] geometry."[20] The excitement generated by projective geometry as a research field is reflected in Smith's elaboration of these interests.

> So great (we might contend) have been the triumphs achieved in [these fields] . . . so vast and so inviting has been the field thus thrown open to research,—that we do well to spend our time and our labour upon a country which has, we might say, been 'prospected' for us, and in which we know beforehand that we cannot fail to obtain results which will repay our trouble.[21]

As important for the attraction of projective geometry as the wealth of new results being produced was the methodology being used to generate these theorems. The new geometry was not perceived as a classically perfect mathematical system built up from clearly defined first principles; rather it was seen in terms Ellis had laid out as early as 1852, when he testified before the Royal Commission that "the method of which it [projective geometry] makes so much use, namely, the generation and transformation of figures by ideal motion, is more natural and philosophical than the (so to speak) rigid geometry to which our attention has been confined."[22] Projective geometry was presented as the new naturalistic study of space which was generated by a scientific methodology of careful generalization and induction. In this way it could be pursued as an integral part of the progressive scientific vision being propounded by a new generation of post-Darwinian scientific publicists.

There were two aspects of projective knowledge which made it particularly comfortable for many young scientists in the final decades of the century. The first was that it was essentially a descriptive study ultimately grounded, like Euclid's, in a concept of space.

---

[20] H. J. Stephen Smith, "On the Present State and Prospects of Some Branches of Pure Mathematics" (Read Nov. 9, 1876), *Proceedings of the London Mathematical Society*, 8 (1876–77):9.

[21] Ibid. pp. 7–8.

[22] Great Britain, Parliament, *Parliamentary Papers, 1852–53*, vol. 44 (*Reports*, vol. 5), "Report of her Majesty's Commissioners appointed to inquire into the State, Discipline, Studies and Revenues of the University and Colleges of Cambridge: together with the Evidence and an Appendix," from "Evidence," p. 224.

Unlike Euclidean geometry, however, projective methods were seen as scientific, closely akin to the inductive generalizing techniques of the natural sciences. Thus projective geometry combined a descriptive approach to geometry with an inductive methodology. This combination was particularly attractive in late-century England. Its form and function can be seen in more detail in English mathematicians' occasional attempts to interpret their subject for wide audiences.

The descriptive emphasis is ubiquitous in English writings about geometry until the end of the century. In this view geometry was the study of a space which existed independently of the mathematical systems used to describe it. For most, mathematical space was not identical to physical space, but was mediated by a neo-Platonic world of ideas. This ideal world was the object of mathematical investigation. This is the view Sylvester espoused in his 1869 address, where he described mathematics as involving

> continually renewed introspection of that inner world of thought of which the phenomena are as varied and require as close attention to discern as those of the outer physical world—to which the inner one in each individual man may, I think, be conceived to stand in somewhat the same general relation of correspondence as a shadow to the object from which it is projected, or as the hollow palm of one hand to the closed fist which it grasps of the other.[23]

It is also the view Arthur Cayley developed fourteen years later in his 1883 address to the BAAS. "I think," Cayley wrote, "it may be at once conceded that the truths of geometry are truths precisely because they relate to and express the properties of what Mill calls 'purely imaginary objects:' . . . I would myself say that the purely imaginary objects are the only realities . . . in regard to which the corresponding physical objects are as the shadows in the cave."[24]

The descriptive aspect of their view of mathematics differentiated these post-Darwinian mathematicians from the analytic approach of the metric non-Euclidean geometers, and tied them closely to the

---

[23]Sylvester, "Plea," (1869), p. 237.
[24]Arthur Cayley, "Presidential Address," *Report of the Fifty-third Meeting of the BAAS Held at Southport in September 1883*, (London: John Murray, 1884), p. 7.

traditional views espoused by Whewell in 1848. The second theme which permeates post-Darwinian writings, however, is inductive. This emphasis distinguished the new generation from its predecessors. The traditional view of mathematics had stressed that results were generated by a deductive form of reasoning which was based on the essential principles of this ideal space. However, the image of a mathematical world of ideal objects lent itself equally well to a quasi-inductive interpretation of method. As the century progressed and the pursuit of the natural sciences was vociferously championed, this inductive view became more prevalent.

Often the inductive view is only subtly expressed in English writings about geometry. It can be seen, for example, lurking behind Smith's geological description of mathematics wherein different fields are countries with lodes of ore to be mined. It also shines through the memorable summarizing passage of Cayley's 1883 address in which he drew an analogy between mathematics and quintessentially observational natural history.

> It is difficult to give an idea of the vast extent of modern mathematics. This word 'extent' is not the right one: I mean extent crowded with beautiful detail—not an extent of mere uniformity such as an objectless plain, but of a tract of beautiful country seen at first in the distance, but which will bear to be rambled through and studied in every detail of hillside and valley, stream, rock, wood, and flower.[25]

What seems almost a subconscious reflection of societal thinking in Smith and Cayley was an expressly developed view of mathematics among a significant group of their contemporaries. In his *Logic,* John Stuart Mill had developed a natural-scientific interpretation of mathematical method at some length as part of his answer to Whewell's more deductively oriented philosophy. Mill was not highly sophisticated mathematically, and on this point his philosophical structure was highly vulnerable.[26] The spirit of his attempt was preserved, however, in the basic orientation of many of England's pro-

---

[25] Ibid., p. 25.
[26] See, for example, William Stanley Jevons, "John Stuart Mill's Philosophy Tested," *Contemporary Review,* 31 (1877–78):167–82, 256–75.

gressive mathematicians. Other younger scholars were more direct than the Oxbridge-established Cayley and Smith were in the connections they drew between mathematical method and that of the natural sciences. Thus, for example, in 1883 Olaus Henrici, who was then professor of mathematics at University College, London, wrote,

> It is the inspection of figures which is of the greatest importance in geometry. It is hereby of little consequence whether the figures are seen by the physical eye or only mentally: because the conception of that space in which we perceive everything and without which we can perceive nothing . . . is built up in our minds through many generations in conformity with sensual impressions.[27]

Henrici's equation of direct physical experience with mathematical experience is concretely reflected in the course of teaching he introduced to University College. In 1870, he succeeded Hirst to the Chair of Pure Mathematics. During his tenure there, Henrici introduced a course in "Modern Geometry and Graphical Statics," which he described as follows in the 1876-77 "College Calendar."

> In this Course Geometry will be treated by modern [projective] as distinguished from Euclidean methods, . . . [the lectures] will supplement those on Coordinate Geometry; and all students who desire to become acquainted with the Theory of Conics and Quadric surfaces, and with the higher branches of Geometry, are strongly recommended to attend them. One of the great advantages of the purely geometrical methods is that all operations are performed by constructions, mostly in three dimensions. Thus the Student learns to realize figures in space, whilst in Coordinate Geometry the geometrical meaning of the algebraical operations is too easily lost sight of.
>   The full benefit of this, however, can be obtained only by applying practically on the drawing-board the theorems and methods given in the lectures to the solution of geometrical problems. Students attending this class are therefore strongly recommended to join the

---

[27] Olaus Henrici, "Presidential Address: Mathematics and Physics Section," *Report of the Fifty-third Meeting of the BAAS Held at Southport in September 1883* (London: John Murray, 1884), pp. 393–400; reprinted in *Nature*, 28(1883):497.

Class of *Geometrical Drawing* which will be conducted throughout in connexion with it.

In order still to further promote this object facilities are provided in the Work-room for the construction of Geometrical models.[28]

The approach to geometrical study here proposed by Henrici was subtly different from the kind of teaching which had been pursued by his illustrious predecessor, Augustus De Morgan. De Morgan's approach had been one in which "a thorough comprehension and mental assimilation of great principles far outweighed in importance any mere analytical dexterity in the application of half-understood principles to particular cases."[29] Henrici's ultimate aim was also to generate an understanding of geometrical principles. He approached this goal from a different angle, however. In his view an understanding of such principles was built up inductively through the direct study of geometrical constructions, rather than first learned intellectually and then applied to individual cases. It is this inductive point of view which was embodied in projective geometry.

The inductive emphasis in projective geometry which linked it methodologically to the natural sciences made the subject significantly different from classical geometry. The contrast is evident, for example, in the "Mathematics" article of the ninth edition of the *Encyclopaedia Britannica:*

> In somewhat sharp contrast with the Grecian geometry . . . stand the different varieties of modern geometry—which aims at greater generality in its definitions, pays less explicit attention to logical form, but arranges geometrical propositions as much as possible in the natural order of development or discovery, and above all makes extensive use of the principle of continuity.[30]

This inductive orientation stopped short of Helmholtz's brand of empiricism, however. Experiments, both in thought and laboratory, were advocated as a method for exploring the rocks and rills of the

---

[28] University College, London, *College Calendar*, 1876–77, p. 31.

[29] Sophia Elizabeth De Morgan, *Memoir of Augustus De Morgan* (London: Longmans, Green and Co., 1882), p. 100.

[30] *Encyclopaedia Britannica*, 9th edition (1883), s.v. "Mathematics," by George Chrystal.

newly recognized space more fundamental than that explored by
Euclid. However, they were not generally recognized as the basis for
defining that space. To quote from Arthur Cayley, "the general
opinion has been and is that it is indeed by experience that we arrive
at the truths of mathematics, but that experience is not their proper
foundation: the mind itself contributes something."[31]

The essentially inductive point of view maintained by Britain's
projective enthusiasts can be clearly seen at the very heart of their
work, in their continued conviction that the principle of continuity
provided an adequate foundation for the new study. In the decades
since it had been so sharply criticized by Cauchy there had been
successful continental attempts, notably in Karl von Staudt's theory
of "throws," to recast the foundations of projective geometry. However, these attempts were irrelevant for England's projective geometers and, for the most part, they ignored them. The first elementary
English treatment of the subject, published in 1865, rested firmly
on the principle of continuity which was stated in the following eminently vague and inductive manner.

> Let a figure be conceived to undergo a certain *continuous* variation,
> and let some *general* property concerning it be granted as true, so
> long as the variation is confined *within* certain limits; then the same
> property will belong to *all* the successive states of the figure (that is,
> all which admit of the property being expressed), the enunciation
> being *modified* (occasionally) according to known rules.[32]

Almost thirty years later, Charles Taylor devoted more than thirty
pages of a "Prolegomena" to his book, *An Introduction to the Ancient
and Modern Geometry of Conics*, to tracing the triumphant development of the principle from Kepler through Boscovich to the nineteenth-century geometers.[33] Cauchy's strictures against the impre-
cise, inductive character of the principle carried little weight in post-
Darwinian England. The research ideal espoused by England's most

---

[31] Cayley, "Address," (1883), p. 5.

[32] John Mulcahy, *Principles of Modern Geometry with Numerous Applications to Plane and Spherical Figures*, (Dublin: Hodges and Smith, 1852), p. 96.

[33] Charles Taylor, *An Introduction to the Ancient and Modern Geometry of Conics*, (Cambridge: Deighton, Bell and Co., 1881).

vociferous scientific advocates was firmly grounded in an inductive interpretation of scientific methodology. For its most enthusiastic supporters, projective geometry was attractive precisely because it brought mathematics into the fold of this inductive approach. In addition, the appeal of the subject was even more powerful after Felix Klein brought the study of non-Euclidean geometry firmly under the umbrella of projective geometry. This development and its implications will be the subject of the next section.

THE PROJECTIVE INTERPRETATION OF NON-EUCLIDEAN GEOMETRY

Almost as soon as non-Euclidean geometry was introduced into England, it was interpreted as a part of the growing field of projective geometry. The mathematical theory for treating non-Euclidean geometries projectively was developed by the German mathematician Felix Klein in papers published in 1871 and 1873. By the end of that decade a significant group of British mathematicians, including Robert Ball, William Clifford, Homersham Cox and Arthur Buchheim, were actively pursuing non-Euclidean geometry from a projective point of view.

Technically, linking non-Euclidean with projective geometries required a mathematical elaboration of Cayley's method of introducing a distance function, or metric, into projective space. In his "Sixth Memoir," as he was developing his model of the distance function in the projective plane, Cayley had pointed out that if the Absolute were interpreted as a degenerate conic—the straight line—the distance relations generated would be those of a flat Euclidean plane. Cayley noted, however, that if it were an imaginary conic, for example the conic described by the equation $x^2 + y^2 = -1$, the relations generated would not be Euclidean; they would exhibit the same properties as the distance relations on a sphere whose antipodal points were identified.

Thus, Cayley explicitly toyed with a two-dimensional non-Euclidean geometry by changing the properties of the distance function he had defined in projective space. There is no further discussion of this point in Cayley's paper, however. His total blindness to the possibility or interest of analogously creating a three-dimensional non-Euclidean geometry from projective space is indicative of his total innocence of non-Euclidean speculation at this time. After the 1860s, however, when synthetic and metric non-Euclidean ideas were flooding both England and the continent, the German mathematician Felix Klein used Cayley's basic approach to generate explicitly non-Euclidean spaces from projective space. Klein briefly presented his projective interpretation in a paper entitled "Ueber die sogenannte Nicht-Euklidische Geometrie," which was published in the *Nachrichten von der Königlichen Gesellschaft der Wissenschaften zu Göttingen* in 1871. He then developed his ideas more fully in two additional papers of the same title, which were published in *Mathematische Annalen*[34]

Klein's projective interpretation of non-Euclidean geometry essentially rested on the theory of distance Cayley had produced for projective space.[35] Klein explicitly developed Cayley's observation that a different distance relation would be generated if the form of the Absolute were changed, into a theory which encompassed the major non-Euclidean geometries discussed by Helmholtz. The Absolute, Klein noted, could take the form of any conic section. In a plane, Euclidean spatial relations would be generated if it were a straight line; for Euclidean three-dimensional space, it would be a plane. But the Absolute could equally take hyperbolic or elliptic forms, which would generate geometries corresponding to Riemann's negative or positive curvatures, respectively. Klein classified the new geometries according to the forms of their Absolutes as "elliptic," "hyperbolic" or "parabolic."

With his projective interpretation of non-Euclidean geometries, Klein provided an approach to the novel geometric systems which was enthusiastically embraced by the English mathematical community. Although his papers were not translated directly into

---

[34] Felix Klein, "Ueber die sogenannte Nicht-Euklidische Geometrie," *Nachrichten von der Königlichen Gesellschaft der Wissenschaften zu Göttingen,* 17 (1871):419–33; Felix Klein, "Ueber die sogenannte Nicht-Euklidische Geometrie," *Mathematische Annalen,* 4 (1871):573–625; Ibid., 6 (1873):112–45.

[35] The only difference was that whereas Cayley used a cosine function to generate distances, Klein used a logarithimic function which was easier to manipulate. Klein's logarithmic function is the one which tends to be used today.

English, numerous references to them in the literature testify to their having been widely recognized. Furthermore, in this case the communication between the German and the English communities was direct; Klein visited England in 1873, and at the Bradford meeting of the BAAS, he discussed his ideas directly with Britain's leading mathematicians. By 1879, Robert S. Ball opened his basic treatment of non-Euclidean geometry with a reference to Cayley's "Sixth Memoir," Klein's paper of 1871 and a subsequent paper by Lindemann. He explained his choice as follows:

> The three works here named are the more conspicuous members of a considerable body of literature in which a remarkable mathematical theory has been developed. The list of writers on the different departments of this theory could be greatly extended, and it would be found to contain the names of many of the most eminent mathematicians, including Gauss, Riemann, Clebsch, and Helmholtz. Although Gauss in one way and Cayley in another are undoubtedly the founders of the Non-Euclidean Geometry, yet it is in Klein's Memoir . . . that the systematic development of the theory is to be found. The subject is there treated with singular elegance and completeness.[36]

Helmholtz, Riemann and the whole metrical approach to the subject play noticeably minor roles in this summary: their work is seen as having been effectively superseded by Klein's.

Ball was not alone in perceiving projective non-Euclidean geometry as the apex of non-Euclidean development. His interest in non-Euclidean geometry stemmed from an interest in the analytical framework of dynamics in non-Euclidean spaces. This interest was one he shared with men like Robert Heath, Clifford, Cox and Buchheim. These men were the most active English developers of non-Euclidean geometry during the 1880s. All of them approached the subject from an essentially projective as opposed to a metric point of view.[37]

---

[36] Robert S. Ball, "On the Non-Euclidean Geometry," *Hermathena*, 3 (1879):500.

[37] For a further treatment of the work of these men, see Renatus Ziegler, *Die Geschichte der Geometrischen Mechanik im 19 Jahrhundert*, (Stuttgart: Franz Steiner Verlag Weisbaden, 1985).

In retrospect, some historians of mathematics have seen this focus on projective geometry as bizarre. It has led Morris Kline, for example, to comment somewhat desperately: "Mathematicians can be readily carried away by their enthusiasms."[38] However, at least in England, the interest in projective geometry is not hard to understand within the larger institutional and epistemological context in which non-Euclidean geometry was received and pursued in the second half of the nineteenth century. Both the synthetic and metric interpretations of non-Euclidean geometry were highly uncomfortable for English mathematicians of the 1860s and 1870s. The projective one was less so. The differences may seem rather subtle, but they were crucial to the way geometrical study was conceived and pursued in late-Victorian England.

The major difference between the projective and metric interpretations of non-Euclidean geometry lay in their scope. When Riemann defined distance he did it locally, examining how the structure of the entire space could be built up from the properties of small parts of it. Klein's projective approach, on the other hand, was global. His distance function depended on the macroscopically given form of the Absolute, not on local properties. In Klein's interpretation, these local properties were dictated by the whole spatial structure; they were fixed in the form of the Absolute, rather than vice versa.

This difference can be illustrated by the fate in the projective interpretation of Riemann's suggestion that there could be spaces of non-constant curvature. Such spaces, where the curvature might vary from point to point, were possible if space were approached as the sum total of its local properties. They could not be projectively interpreted, however, because the form of the Absolute determined a fixed distance relationship among all points in a projectively embedded metric space. By interpreting non-Euclidean geometry projectively, then, mathematicians were ruling out some of the flexibility possible in Riemann's view.

This narrowing of possibility was mathematically easy to accept in the late nineteenth century because it was generally considered to

---

[38] Morris Kline, *Mathematical Thought from Ancient to Modern Times*, (New York: Oxford University Press, 1972), p. 923.

be necessary for any real geometries. Helmholtz's psychological interpretation of Riemann's work persuaded his contemporaries that the real, experiential basis for Riemann's local distance was rigid body motions. The properties of such motions implied that space must be of constant curvature. For this reason, Riemann's speculations about non-constant curvature were almost universally rejected from inside the metric school, even before Klein left them out of his projective interpretation.

Although in late nineteenth-century practice there was virtually no disagreement about spaces with non-constant curvature, the difference in scope of the metric and projective viewpoints significantly distinguished them from a philosophical perspective. The fine focus of the metric geometers was presented as the cutting edge of a radically empirical philosophy which threatened to undermine the conceptual view of geometry. Instead of viewing geometry as the exploration of an autonomous spatial concept, the metric geometers asserted it was merely a summary of contingent experience. As has been shown above, this interpretation threatened both the philosophical and institutional position of geometry in nineteenth-century England. From a philosophical standpoint it threatened the notion of necessary truth which was central to natural theology. Furthermore, as Huxley's argument suggests, alternative interpretations of geometry's truth value could be used to argue against the intellectual value of the study at all.

With its global approach, on the other hand, the projective interpretation allowed the conceptual view of geometry to remain essentially intact. It is true that what had hitherto been accepted as the spatial concept—Euclidean space—was replaced by projective space. But this did not change the basic epistemological position of the spatial concept which was still an autonomous intellectual entity, independent of the mathematical systems describing it. Therefore, interpreting non-Euclidean geometries within a projective framework significantly tempered the radical claims which had been made within the metric tradition.

Projectively interpreting non-Euclidean geometry did not completely settle questions about the nature of geometry, however. The kind of epistemological issues Helmholtz and Riemann had raised were sidestepped with the new approach. Nonetheless, there was another, closely related problem which lay at the base of *all* projective study. When it was projectively interpreted, non-Euclidean geometry became involved in the conceptual problem surrounding imaginary and infinite points.

THE CONCEPTUAL FOUNDATIONS OF PROJECTIVE GEOMETRY

The foundational difficulty with projective geometry existed before non-Euclidean geometries were incorporated into its structure. It is a problem which can be seen lurking in Sylvester's 1869 address, though he brushed over it so lightly as to render it almost invisible. In that address, Sylvester advocated "early familiarization" not only with the "doctrine of the imaginary" but also with that of "the inconceivable." This is a peculiar statement because in the accepted interpretation of the nature of geometry it was just the crystal clear conceivability of its elements which was the key to its peculiar form of truth.

Projective geometry, however, seemed to contain inconceivable elements. The points of intersection of two circles which neither touch nor cut each other have no place in the usual image of space. The point of intersection of a pair of parallel lines might be suggested by looking down a railroad track, but anyone who has traveled down such a track knows it to be an illusion. Imaginary points and points at infinity were not clearly conceivable. Yet these points were an essential part of projective geometry's success in providing geometrical generality to match that of analysis. In addition, they were essential to the Cayley/Klein theory of distance. The whole theory of the Absolute, which provided the additional points for the cross ratio definition of distance, was composed of these evidently inconceivable points.

Because of their conceptually oriented view of geometry, England's mathematicians were forced to specifically address the conceptual problems raised by projective space. Although conceptual projective problems were not presented as urgently as were those which surrounded metric or synthetic non-Euclidean geometries, they proved to be equally intractable. Those who considered the imaginary and infinite points late in the 1880s seemed equally unsure about how to conceive of them as did those who first considered them twenty years before.

From an early date, one can find explicit statements recognizing the basic conceptual issues which surrounded projective geometry's imaginary and infinite points. One of the earliest, as well as the most colorful, of these is in an 1866 article by Ferdinand Wolff. "What most distinguishes Ancient Geometry from the Modern," wrote Wolff,

> is that the idea of infinity and evanescence was incompatible with the plastic mind of the Greeks. . . . If you had told them that parallel lines meet at infinity, Euclid would have been the first to cry out, 'By Pluto, do they meet then in the kingdom of the shades, as they cannot meet in this world *though indefinitely produced'* . . . And as for circles if Poncelet himself had expounded in the face of Greece his theory of all the circles cutting in the same two points at infinity, Pythagoras . . . would have pounced upon him exclaiming in despair, 'At two points indeed, and if the one lies in the Elysian fields and the other in the infernal regions, by Cerberus how can I ever hope to get at them both?'[39]

Wolff's exuberant treatment located the problem, but he did not offer a solution.

Others did try to compose answers to the Greeks' questions, however. English mathematicians were often in a position to examine the nature of projective geometry in introductions or conclusions to articles, in encyclopedias and textbooks, in public addresses to their peers or wider scientific audiences. Similar issues were also addressed in their professional papers themselves. From these sources one can construct a surprisingly consistent interpretation of the nature of geometry. From within this view emerges a solution to the problem of the conceptual status of infinite and imaginary

---

[39] Ferdinand Wolff, "Critical Examination of Euclid's First Principles Compared to Those of Modern Geometry, Ancient and Modern Analysis," *The Quarterly Journal of Pure and Applied Mathematics,* 8 (1866–67):307–08.

points which serves to underline and reemphasize the inductive, scientific orientation which was attendant on projective geometry.

The quasi-inductive and scientific solution to projective geometry's foundational problems is illustrated by Henrici's treatment of points at infinity in the article "Geometry" in the *Encyclopaedia Britannica*. There Henrici posed a problem which he saw as suggesting that there be several infinite points at the end of two parallel lines rather than one.[40] He resolved it as follows:

> The difficulty in which we are thus involved is due to the fact that we try to reason about infinity as if we, with our finite capabilities, could comprehend the infinite. To overcome this difficulty, we may say that all points at infinity in a line *appear* to us as one, and may be replaced by a single 'ideal' point, just as all points in a fixed star—which is not at an infinite, only at a great distance—cannot be distinguished by us and to beings on earth count as one.[41]

Projective geometry *did* have the qualities of an observational science, and some of its pitfalls as well. As the astronomer had to grope past the limits of his perception for his understanding of celestial bodies, so projective geometers groped after an understanding of elusive geometrical points. That the understanding had not been reached did not lead English mathematicians to conclude that interpretation was impossible, that geometry was a formal rather than a spatial study. The history of science, they noted, contained many examples of theoretical entities discovered by reason before being recognized by observation. Adams and Leverier, for example, had "discovered" Neptune and even calculated its orbit before it had been observed. In the same way, geometrical reasoning pointed to the presence of imaginary points before their actual position was clear.[42]

---

[40] His problem is stated as follows: "*Every* line through S which joins it to any point at an infinite distance in *p* is parallel to *p*. But by Euclid's 12th axiom there is but one line parallel to *p* through S." *Encyclopaedia Britannica*, 9th edition, (1883), s. v. "Geometry," by Olaus Henrici.

[41] Ibid.

[42] For a development of this kind of argument, see Karl Pearson, *The Grammar of Science*, (London: Walter Scott, Ltd., 1895), Chap. V, "Space and Time."

This did *not* mean that the interpretation was unattainable. In the long run Neptune *had* been observed, and similarly someone would find a way to interpret the imaginary points demanded by reason. In his 1878 Presidential Address to the BAAS, William Spottiswoode remarked:

> it does not . . . follow that upon a more enlarged basis the formulae are incapable of interpretation; on the contrary, the difficulty at which we have arrived indicates that there must be some more comprehensive statement of the problem which will include cases impossible in the more limited, but possible in the wider view of the subject.[43]

He continued comfortingly:

> even if . . . the most comprehensive statement of the problem of which we are at present capable fails to give an actual representation of these quantities; . . . it still does not follow that we may not at some future time find a law which will endow them with reality, nor that in the meantime we need hesitate to employ them, in accordance with the great principle of continuity, for bringing out correct results.[44]

Spottiswoode's position and that of England's other naturalistic mathematicians is open-ended—the interpretation of imaginary circular points remained elusive. However, their reality was not seriously questioned. England's progressive mathematicians pursued the subject unshaken in the faith that eventually an interpretation would be found.

A similarly open-ended approach to projective geometry's conceptual problems is to be found in the writings of Cayley. Cayley held a singularly promising position from which to reevaluate the foundations of mathematics, specifically geometry, when he gave his Presidential Address to the BAAS in 1883. Almost twenty-five years before, with his assertion that projective geometry is *all* geometry, Cayley had set the stage for approaching projective space as a con-

---

[43] William Spottiswoode, "Presidential Address," *Report of the Forty-eighth Meeting of the BAAS held at Dublin in August 1878*, (London: John Murray, 1879), p. 18.
[44] Ibid., p. 19.

ceivable concept. Klein's projective interpretation of non-Euclidean geometries relied essentially on this insight.

As was pointed out above, Cayley drew a distinctly Platonic picture of mathematics in his Southport address. Even as he did so, however, he conceded that some important pressures had been created by new geometrical developments. The problems did not arise from non-Euclidean geometry *per se,* though. Cayley dismissed the metric challenge to geometrical truth with an argument strictly separating necessary geometrical from contingent experiential truth, which strikingly resembles the one Jevons had mounted against Helmholtz thirteen years before.

The ontological status of imaginary points, both in geometry and in analysis, was a problem with which he was explicitly less comfortable, however. "The notion of a negative magnitude has become quite a familiar one, . . . " he wrote.

> But it is far otherwise with the notion which is really the fundamental one (and I cannot too strongly emphasise the assertion) underlying and pervading the whole of modern analysis and geometry, that of imaginary magnitude in analysis and of imaginary space . . . in geometry: I use in each case the word imaginary as including real. This has not been, so far as I am aware, a subject of philosophical discussion or enquiry. . . . [C]onsidering the prominent position which the notion occupies—say even that the conclusion were that the notion belongs to mere technical mathematics, or has reference to nonentities in regard to which no science is possible, still it seems to me that (as a subject of philosophical discussion) the notion ought not to be thus ignored; it should at least be shown that there is a right to ignore it.[45]

Having recognized the ontological problems raised by projective geometry, however, Cayley moved away from them. He did not offer a solution but rather exhorted a different group, the philosophers, to solve it for him.

They did not, however. Perhaps in part because projective geometry and the themes it encompassed were too esoteric to be known by the philosophical community, there was no philosophical discus-

---

[45] Cayley, "Address" (1883):8–9.

sion of the issues Cayley felt were raised by projective geometry until late in the final decade of the century. In his Southport address, Cayley refrained from the blatant naturalistic arguments of Spottiswoode or Henrici but offered nothing substantial in their stead. So, even in the hands of its most illustrious developer, it seems that projective geometry remained foundationally flawed.

It is tempting to dismiss the whole genre of popular addresses as unreliable witnesses to mathematicians' real views. However, there is little evidence to suggest that within their own ranks England's mathematicians advocated a significantly different position about the nature of geometry than that put forward in popular addresses. By the second half of the nineteenth century, a significant community had developed which pursued mathematics independent of the old justificatory framework of the liberal education. Although the development of paying jobs which would support mathematical research outside of the unifying structure of education lagged considerably, a recognizable community with a focused mathematical universe of discourse had developed by the early 1860s. The boundaries of this community can be crudely traced by the purview of journals like the *Quarterly Journal of Pure and Applied Mathematics,* *The Messenger of Mathematics* or by membership in specifically mathematical organizations like the London Mathematical Society.

In this mathematical universe of discourse, the discussion of the conceivability of projective geometry did not revolve around the ontological status of imaginary points and points at infinity. Although their popular utterances reveal that England's mathematicians recognized these points as a conceptual challenge to which they did not have a clear response, these problems do not arise in the context of their mathematical work. What did arise in the latter context were closely related questions about Cayley's definition of distance and the mutual relationship of Euclidean, non-Euclidean and projective geometries.

The specific point of strain lay in the concept of distance which Cayley had introduced and Klein had modified. In his "Sixth Memoir," Cayley claimed to have established the notion of distance on "purely descriptive [projective] principles."[46] He took pains to dem-

---

[46] Cayley, "Sixth Memoir," (1859) p. 561.

onstrate that the new, anharmonic definition he developed displayed the same algebraic properties as the ordinary definition of distance. In the conclusion to the "Memoir," however, Cayley went beyond this formal statement. In claiming that projective geometry "was *all* geometry," he was asserting the real, conceptual primacy of his projective notion of distance, including the somewhat peculiar implication that distance was a relationship among four points rather than two: "the theory in effect is, that the metrical properties of a figure are not the properties of the figure considered *per se* apart from everything else, but its properties when considered in connexion with another figure, viz. the conic termed the Absolute."[47]

In the seventh and eighth decades of the century, Cayley's projective interpretation of non-Euclidean geometries was of particular interest to the group who were exploring the analytical framework for dynamics in non-Euclidean spaces. Interpreting Euclidean space as the limiting case of elliptic space, they directed their researches at developing the most general forms of dynamic theories. The Cayley/Klein definition of distance was critical to this enterprise.

They recognized, however, that there was an important conceptual problem with Cayley's theory of distance, which is indicated in a paper by Ball published in 1879. There Ball presented the Cayley/Klein interpretation of metrical geometries as follows:

> To take the first step in the exposition of this theory, it is necessary to replace our ordinary conception of distance, or rather of *the mode in which distance is measured* by a more general conception.

He then introduced the fundamental quadric—Cayley's Absolute—and the logarithmic definition of distance. "It cannot be denied," he continued,

> that there *appears* to be something arbitrary in this definition when read for the first time. But as the reader proceeds he will find that it is at all events *plausible*, even though he may not go so far as to agree

---

[47] Ibid., p. 592. The full quotation is on p. 130.

with those who consider that any other conception of distance is imperfect.[48]

Here Ball is clearly concerned as much with the conceptual issues raised by the Cayley/Klein definition of distance as with their mathematical usefulness. Throughout the article he struggled not only to develop a mathematical structure, but also to reconcile intuitive spatial ideas with those which were being mathematically generated in projective geometry.

The conceptual difficulty with the four-point projective definition of distance was exacerbated by a circularity which seemed to lurk at its core. The distance between two points was defined as a function of their cross ratio with points on the Absolute, but this ratio itself was composed of distances measured between points. Thus, projective space seemed somehow to come out of metric space only to move back into it with a more complicated definition of distance.

Looking back on the "Sixth Memoir" when it was republished in his *Mathematical Papers* in 1889, Cayley appended a "Note" reconsidering his original theory of distance. When he originally wrote the "Memoir," Cayley commented, he had considered coordinates as "an assumed fundamental notion, not requiring or admitting of explanation."[49] From this standpoint, distance was a conventionally defined function of these coordinates, and the various non-Euclidean metrics were formed by changing these conventionally defined functions.

The mathematicians who pursued Cayley's ideas in dynamics usually adopted uncritically Cayley's view that the measure of distance in space was conventionally defined. Spatial coordinates were taken as primary, and distance was defined from these coordinates. The differences between Euclidean and non-Euclidean spaces arose as the result of arbitrarily changing the coordinate function which measured the distance between spatial points. To quote from Cox:

> In Prof. Cayley's [also Cox's] method of considering the subject it is not the properties of space but our manner of measuring it that has

---

[48] Ball, "Non-Euclidean," (1879):502.

[49] Arthur Cayley, "Notes and References," *The Collected Mathematical Papers of Arthur Cayley*, 11 vols. (Cambridge: University Press, 1893), 2, p. 605.

changed. By measuring ordinary space in a particular manner we obtain the same relations as those of hyperbolic space.[50]

Cox's presentation stresses the purely conventional side of Cayley's interpretation of distance.

Thirty years later, however, when he appended his "Note," Cayley recognized that other interpretations were possible. One possibility was that "[coordinates] might be regarded as mere numerical values, attached arbitrarily to the point, in such wise that for any given point the ratio $x:y$ has a determinate numerical value, and that to any given numerical value of $x:y$ there corresponds a single point." In this approach the status of distance and the coordinates was reversed. Distance was treated as fundamental. It existed prior to the coordinates, which were arbitrarily assigned values, limited only by the necessity that the distance relation be true.

The second alternative Cayley considered was von Staudt's, where "the anharmonic [cross] ratio of any four points, has independently of any notion of distance the fundamental properties of a numerical magnitude, viz. any two such ratios have a sum and also a product, such sum and product being each of them a like ratio of four points determinable by purely descriptive constructions." Cayley did not develop these ideas further, but he was clearly wary of this approach. He feared that somehow the argument was circular, "since the construction for the product of the two ratios, is in effect the assumption of the relation, dist PQ + dist QR = dist PR."

Cayley closed his discussion on an ambiguous note. Having briefly presented his original view and these two alternative approaches to the concept of distance, he did not choose among them. Even in criticizing von Staudt's approach his language is highly circumspect. "There is at least the appearance of arguing in a circle,"[51] he wrote, but he did not take a stand as to whether the appearance was real or not.

Cayley's failure to resolve the issues raised in this "Note" points

---

[50] Homersham Cox, "Homogeneous Co-ordinates in Imaginary Geometry and their Application to Systems of Forces," *The Quarterly Journal of Pure and Applied Mathematics,* 18 (1881):184.

[51] Cayley, "Notes," (1893), p. 605.

to the difficulty of the issues projective geometry raised for the English insistence that conceptual clarity was the ultimate ground of geometrical development. Within this view, mathematical entities described conceptual realities. Therefore, if it were valid, Cayley's projective theory of distance must be susceptible to conceptual analysis. Yet Cayley himself was unable to cut the Gordian knot of conceptual confusion which surrounded it.

Within a decade of being first widely introduced into England, non-Euclidean geometry was reinterpreted within a projective framework. This reinterpretation served to integrate the new geometries in a way that was highly attractive for those who were defending the value of mathematics within a new, naturalistic approach to knowledge. At the same time, it both defused the radical claims about necessary truth associated with synthetic and metrical non-Euclidean geometries and turned discussion away from abstruse philosophical considerations of the nature of "conception," "imagination," "truth" and so on. Non-Euclidean geometry lost a great many of its disturbing qualities by becoming part of projective geometry.

There were still problems associated with projective geometry, however, which ultimately were very close to the conceptual ones originally connected with non-Euclidean geometry. These problems were more esoteric, and hence seemed less central to the status of mathematics, than those raised by non-Euclidean geometers like Helmholtz. Therefore, they did not draw the same amount of general intellectual and philosophical interest. In their way they were no less intractable, though, and England's geometers were unable to solve them. Their resolution had to wait until the larger social and intellectual configuration which surrounded geometry changed enough to accommodate a radically new vision of the entire enterprise.

The resolution of the conceptual difficulties associated with projective geometry was achieved only with the overthrow of the entire attempt to interpret the foundations of geometry descriptively. The forces involved in this change of outlook, which gathered strength at the turn of the century, extended far beyond geometry itself. The changing position of geometrical study within English culture in the final decades of the century can be best seen by looking closely at another series of developments apparently more social than intellectual. In the context of discussions about teaching elementary geometry, England's mathematicians actively struggled with a number of practical problems which were involved in their descriptive view of mathematical truth. The next chapter will explore the range of intellectual issues which accompanied their attempts to reform elementary geometrical education in England. Although the educational conflicts turned around elementary geometry, rather than the forefront of research where non-Euclidean and projective geometries were being developed, the basic questions being addressed were often strikingly similar. The educational situation provides further insight into the range of influences which contributed to English views of the nature of mathematics late in the century.

Title page of the first English edition of Euclid. *(Courtesy of the John Hay Library at Brown University.)*

CHAPTER 4

Euclid and the English Schoolchild

The development of both non-Euclidean geometries and projective geometry introduced significant strains into the previously unified interpretation of the nature of geometry. It is tempting to see these mathematical developments as necessitating a change in the English cultural interpretation of mathematics. In the arguments of the scientific naturalists like William Clifford, the development of non-Euclidean geometry is presented as forcing those who understood it to abandon their previous belief in the absolute truth of geometry. From this perspective, non-Euclidean developments led inexorably away from the conceptual view of mathematical truth and forced changes in the whole truth structure that was erected on this basis.

The projective interpretation of non-Euclidean geometry significantly complicates this picture, however. A whole layer of negotiable interpretation, which Clifford did not recognize, or at least did not mention, intervenes between the development of mathematical forms and decisions about their meaning or epistemological implications. The differences among the synthetic, metric and projective interpretations of non-Euclidean geometry illustrate how disparate various interpretations of the same mathematical forms can be. In England during the 1860s and 1870s, the projective interpretation of non-Euclidean geometry significantly diminished the compelling power of Clifford's metrically based claims that non-Euclidean developments necessitated epistemological change.

Nevertheless, the case that mathematical development unilaterally forced a change in perceptions of mathematical truth can still be forcefully argued. The late nineteenth-century foundation problems of projective geometry have been seen as leading just as inevitably toward the destruction of the conceptual view and the epistemological structure it supported. In a paper entitled "The Formation of Modern Conceptions of Formal Logic in the Development of Geometry," Ernest Nagel has argued that the conceptual problems raised by projective geometry's infinite and imaginary points forced mathematicians to abandon the whole notion that geometry was descriptive. In his words, the abandonment of the descriptive view of geometry "owes next to nothing to the speculations of professional philosophers and logicians, and is the outcome of technical needs and advances of mathematics proper."[1]

It is undeniable that in the long run, imaginary points and points at infinity were easier to understand outside of the late nineteenth-century interpretation of mathematical truth. Among late nineteenth-century English mathematicians, however, it seems equally clear that a descriptive view of geometry was much more central to their enterprise than was the foundational status of these points. Imaginary and infinite points posed a relatively minor and recondite problem for their conceptual view. To reinterpret the nature of the entire geometrical enterprise in order to deal with such a minor irritation would be to throw the baby out with the bath.

Nonetheless, the conceptual view in which geometrical theorems described the properties of a conceptual space was ultimately abandoned in England at the turn of the century. Certainly, though they did not necessitate change, the kinds of issues posed by non-Euclidean and projective geometries contributed to the power of the more formalistic definitional arguments when they were accepted. At least equally important to this acceptance, however, were changes in the broad intellectual and cultural matrix in which geometry was embedded. In the second half of the century, significant alternatives

---

[1] Ernest Nagel, *Osiris*, 7 (1939):143–44.

to the truth-oriented, mid-century view of knowledge grew ever stronger. Huxley's educational vision, which emphasized the value of inductive investigation over the recognition of absolute, fixed truth, is an example of this kind of alternative. Sylvester's reply to Huxley shows that mathematical study could be construed in such a way that it would be supported within such alternative educational philosophies. This revisionary enterprise required reexamining the nature of mathematics, however, and opened the door for differences of opinion about what comprised the essence of the subject.

Ultimately, all discussions about the nature of geometry are discussions about its foundations, whether they take place at the forefront of mathematical research or on the apparently mundane level of elementary teaching. This is nowhere clearer than in England in the 1870s and 1880s. During this period virtually all of England's mathematical enthusiasts were directly involved in considering the optimal form of elementary geometry texts. The geometrical substance of their discussions was relentlessly elementary; subjects on the forefront of research, like non-Euclidean or projective geometry, were not at issue. Nevertheless, many of the same philosophical issues which faced mathematicians at these cutting edges of geometrical development were also central to these educational matters. In particular, the nature of the conceivable was as central an issue for educators trying to develop student understanding as it was for researchers exploring its farthest reaches. In the educational arena, the societal and cultural pressures on the descriptive view of geometry were being constantly fielded and explicitly negotiated.

This chapter, then, will focus on the late-century English textbook discussions. In the first section, the basic institutional parameters of the controversy will be laid out. These will provide the backdrop for a thematic analysis of some of the major foundational issues involved which will comprise the second section. The third section will consider the work of Edward Travers Dixon, a little known figure who published several works on geometry early in the 1890s. Dixon's work can be seen as the culmination of the textbook debates, illustrating both the power and the limitations of the kinds of thinking about geometry which were generated in this context.

THE DEBATE OVER TEXTS

If one expands the term to encompass its literal meaning, the late-century textbook discussion revolved around the teaching of "non-Euclidean" geometry. In education, however, the issue was not whether Euclid's system adequately described space, but whether his book itself was the best text from which to teach the subject. Throughout the eighteenth and early nineteenth centuries, the *Elements* had been the standard elementary textbook for all English geometry students. Euclid was available in a variety of editions, but ultimately all educational innovation was circumscribed by his towering authority.

In the 1860s, however, this situation was challenged. A number of those involved in mathematical education began to question the adequacy of Euclid's *Elements* as a textbook for schoolboys. Sylvester's 1869 "Plea for the Mathematician," for example, was largely a call to abandon the *Elements* as a text. His request that mathematics be taught with the "life and animation" of the inductive sciences was directly linked to abandoning "our traditional and mediaeval modes of teaching." In this phrase he was referring to teaching from Euclid. "I should rejoice," he declared, "to see . . . Euclid honourably shelved or buried 'deeper than did ever plummet sound' out of the schoolboy's reach."[2]

Since, as Sylvester noted elsewhere in his speech, "there are some who rank Euclid as second in sacredness to the Bible alone, and as one of the advanced outposts of the British Constitution,"[3] his suggestion to replace the Greek with an alternative text was a radical one. It was also timely. The late 1860s witnessed the publication of a number of texts offered as alternatives to Euclid's *Elements*. What began as a trickle in the 1860s was a veritable flood by the 1870s. In the first decade, the pages of Huxley's new journal, *Nature,* fairly bristle with discussions of the optimal form of geometrical teaching.

---

[2] James J. Sylvester, "Presidential Address: Mathematics and Physics Section," *Report of the Thirty-ninth Meeting of the BAAS held at Exeter in August, 1869,* (London: John Murray, 1870), pp. 1–9; reprinted as "A Plea for the Mathematician," *Nature,* 1 (1869–70):261.
[3] Ibid., p. 262.

Text after text was critically reviewed, editorials examined the goals of mathematical teaching, and concerned parents, teachers and students wrote their views in letters.

The positions in the Huxley-Sylvester debate were basically on the periphery of the textbook controversy, where the question was not whether geometry should be taught at all but rather how it should be taught. The challenge Huxley's kind of inductive emphasis posed for the more traditional truth-oriented approach to mathematics was a central feature of the debate, however. His suggestion that the goal of education was to encourage original investigative skills was a central issue around which the textbook discussions turned.

The role differences in educational ideology and goals played in the controversy can be clearly seen in the opening round which was played out between the then aged Augustus De Morgan of University College and James Wilson, the young mathematical master at Rugby. In 1868, hoping to make geometry more palatable for the schoolboy, Wilson published a textbook entitled *Elementary Geometry*. In the "Preface," which Wilson also delivered as a paper before the London Mathematical Society, he directed a barrage of criticism against the educational value of Euclid's *Elements*. De Morgan, who was the founder and first president of the society, promptly responded with a negative review of Wilson's proposed alternative in the *Athenaeum*. Although he was England's most astute critic of Euclid's work, De Morgan unambiguously hailed it as superior to Wilson's. The discussion continued in further letters from each man to the *Athenaeum*'s editor.[4]

The mathematico-philosophical issues in the De Morgan-Wilson exchange continued to echo in the textbook discussions of the subsequent decades; many of them will be considered in the second section of this chapter. For the purposes of this more educationally focused section, their educational ideologies are relevant because

---

[4] James Wilson, *Elementary Geometry*, (London: Macmillan and Co., 1868). The "Preface" to this work was published separately under the title "Euclid as a Text-Book of Elementary Geometry," *The Educational Times and Journal of the College of Preceptors*, 21 (1868):126–28. The review, universally attributed to De Morgan, appeared in *The Athenaeum*, no. 2125 (July 18, 1868):71–73. Wilson replied in *The Athenaeum*, no. 2129 (August 15, 1868):216, and De Morgan again responded in *The Athenaeum*, no. 2130 (August 22, 1868):241–42.

they indicate the major positions around which the textbook discussants grouped themselves in the years which followed.

In his "Preface," which he reprinted in the numerous subsequent editions of *Elementary Geometry*, Wilson charged that Euclid's *Elements* was artifical, invariably syllogistic in form, tediously lengthy in demonstration and, finally, unsuggestive. All of these criticisms can be seen as closely related to and reliant on the first of them, Euclid's artificiality. Euclid, Wilson complained,

> aimed, not at unfolding Geometry as a science, but at shewing on how few axioms and postulates the whole could be made to depend: and he has thus sacrificed, to a great extent, simplicity and naturalness in his demonstrations, without any corresponding gain in grasp or cogency.[5]

The view of geometry that Wilson offered as an alternative to this rigidly ordered one was essentially the same as the one that Sylvester picked up on and advocated a year later in his BAAS address. The Rugby master argued that geometry was the scientific study of space. Its pursuit should involve all of the multifarious reasoning faculties involved in the inductive sciences, not merely formal proof structures. As a textbook, Euclid's *Elements* was inadequate, because it did not achieve this end. "[Euclid's] reasoning is exquisite and profound," Wilson noted, but for the purposes of education "it is too exquisite; it leaves on most men's minds the half-defined impression that all profound reasoning is something far-fetched and artificial, and differing altogether from good clear common sense."[6] In *Elementary Geometry*, Wilson tried to present geometrical theorems in ways which would encourage students to think them out independently. This attempt was presaged on a liberal ideology of intellectual self-sufficiency.

De Morgan's critique of Wilson rested on a more austere and conservative view of the educational benefits to be gained from the study of geometry. He felt that the subject's value lay just in the

---

[5] Wilson, *Elementary Geometry*, (1868), p. v.
[6] Ibid., pp. xi–xii.

sobering effect of its exquisite reasoning on the over-eager young mind.

> Geometry is intended, in education, . . . to [teach] the tricks which reason plays on all but the cautious, *plus* the dangers arising out of caution itself. Let him that thinketh he standeth take heed lest he fall, is the motto on the door; augmented by, Let him that thinketh he falleth be not quite sure he doth not stand both until after very close examination. . . . Such a pause [or examination] is never made by a young student: he is quick to see everything except that he sees too much and too fast. Of all his studies, geometry is the one in which experience may cure him of this nasty habit, if he be properly exercised in the very field of danger.[7]

De Morgan found that Euclid's reasoning, which Wilson had described as "far-fetched and artificial," was perfectly suited to what he saw as the salutary end of tempering trust in the "good clear commonsense" Wilson so admired. The older man advocated the study of Euclid as the perfect palliative for just the kind of exuberant exploration Wilson hoped to encourage.

The essential point Wilson and De Morgan held in common was that the study of geometry was a good way to teach young men how to reason. Their educational disagreement was over the emphasis in this equation; Wilson wanted his liberal view of reason to shape geometry, De Morgan wanted his austere view of geometry to be transferred into other areas of thought. In an important sense, the educational discussions were a tug of war between these two emphases, with the conviction that geometry was a preeminently reasonable study as the connecting rope.

De Morgan died in 1871, but his defense of Euclid was carried on by a strong conservative group long after he faded from the scene. In the writings of men like Charles Dodgson at Oxford or Isaac Todhunter at Cambridge, De Morgan's argument that Euclid was valuable largely *because* of the difficulties of its reasoning was often supplemented by a humanistic one. Within a liberal educa-

---

[7] De Morgan, "Review," (July 18, 1868), p. 72.

tion, many conservatives emphasized, it was not merely geometry which was being taught, it was *Euclid's* geometry. This gave elementary geometry not only a disciplinal but a humanistic interest.[8] When the students learned Euclid, they were initiated into a universe of discourse which had persisted for thousands of years. As Dodgson emphasized in defense of retaining Euclid's ordering of theorems:

> The Propositions have been known by those numbers for two thousand years; they have been referred to, probably, by hundreds of writers . . . and some of them, I.5 and I.47 for instance—'the Asses' Bridge' and 'The Windmill'—are now historical characters, and their nicknames are 'familiar as household words.'[9]

However, the appeal of this kind of historical, humanistic argument was lost on the growing number of Englishmen who were pursuing geometry as preparation for life as an engineer or some other practical career.

In the last half of the century, a strong movement challenged the value of this classical emphasis in education. A growing middle class began to exert increasing pressure for a more practically oriented approach. A major indicator of this pressure can be seen in the movement towards teaching the natural sciences at Cambridge. In the Whewellian mid-century view, such subjects were progressive and therefore should not be taught until permanent studies had been mastered. The growing strength of a different educational ideal, which recognized the value of research-oriented scientific study, can be measured by the success of the Natural Science Tripos. First instituted in 1851, this examination attracted a negligible number of students in its first thirty years. In the next thirty-five years, success on this examination became increasingly linked with careers in medicine and other nonclerical professions. During this period, the number who took it increased more than ninefold. This

---

[8] For a development of this point of view, see William Whewell, *Of a Liberal Education in General*, (London: John W. Parker, 1845).
[9] Charles L. Dodgson, *Euclid and his Modern Rivals*, (London: Macmillan and Co., 1885), p. 11.

extraordinary increase reflects the surge of interest in the middle-class ideal of an education for practical as well as humanistic ends.[10]

Cambridge was following rather than leading the country with these developments. The need for more practical scientific education was pushed on them by an increasingly vociferous middle class. The pressures to which they were responding can be illustrated by Huxley's testimony before a Select Committee on Scientific Instruction in 1868. In response to questions about how he felt a hypothetical son would best be educated for a career in manufacturing—that is, in business—Huxley recommended avoiding an Oxbridge education entirely. "If I intended my son for any branch of manufacture, I should not dream of sending him to the university. I should send him to a good school, and then make him matriculate at the London University, and then I should devote him entirely to scientific pursuits."[11]

The power of Huxley's insistence on the importance of a different ideal of education for professional life was heightened by a widespread fear that England was falling behind its continental neighbors in scientific and technological prowess. The ubiquity of this feeling can be illustrated by the fact that it was even shared by Matthew Arnold, who stated, "In nothing do England and the Continent at the present moment more strikingly differ than in the prominence which is now given to the idea of science there, and the neglect in which this idea still lies here. . . ." Though properly an intellectual agency, he continued, our school system "has done and does nothing to counteract the indisposition to science which is our great intellectual fault."[12] It was basically as a response to this kind of extra-University pressure, generated by a changing professional structure and by scientific xenophobia, that the inductive sciences

---

[10] For a more detailed treatment of this movement, see Roy MacLeod, "The Naturals and Victorian Cambridge," *Oxford Review of Education*, 6 (1980):177–95.
[11] Great Britain, Parliament, *Parliamentary Papers, 1867–68*, vol 15 (*Reports from Committees*, vol. 10), "Report from the Select Committee on Scientific Instruction; with the Proceedings of the Committee, Minutes of Evidence, and Appendix," "Minutes of Evidence," p. 402.
[12] Matthew Arnold, *Higher Schools and Universities in Germany*, 2nd ed., (London: Macmillan and Co., 1892), p. 198.

became a legitimate part of the late-century English educational ideal.

As was shown in the last chapter, these pressures to adjust to a changing image of knowledge were also felt in the mathematical community and had significant repercussions for their pursuit of geometry. They fueled English interest in the practically rooted, quasi-inductive study of projective geometry. In addition to shaping their research interests, these pressures affected their educational views. The role of a mathematician was still a broad one, and most of England's mathematicians were directly involved in educational issues. Many of the progressively minded ones actively tried to promote a new, more liberal approach to their subject in the context of basic education.

In this educational arena, however, reform-minded mathematicians were faced with a difficult problem. The conservative defense of Euclid and his *Elements* was bolstered a great deal by the form of the prevailing institutional structure. In England, not only educational but often professional position and promotion in the army, navy, civil service and so on were routinely judged by standardized examinations. In geometry, these examinations assumed a thorough knowledge of specific theorems, proofs, orderings and so forth from Euclid's *Elements*. Rejecting Euclid as the basic text required drastically revising the examinations and developing a new, equally clear-cut and universal standard by which students could be judged.

This was a tall order but one England's reform-minded educators tried to meet head-on. In 1870, a meeting of thirty-six headmasters of public schools overwhelmingly passed a resolution calling for a general reevaluation of Euclid as a textbook; in the following year a special committee of the BAAS, including Cayley, Clifford, Hirst, H. J. S. Smith, Salmon and Sylvester was formed to look into the matter. The widespread concern culminated in the formation of the Association for the Improvement of Geometrical Teaching, the AIGT, which held its first meeting at University College in January of 1871.

The basic agenda for the early years of the AIGT was established at the first meeting. At that time, Hirst, who was the group's first president, maintained that "an improved treatise on Geometry, for educational purposes, [is] an admitted national want." He directed the attention of the Association "to the surest means of meeting that want satisfactorily."[13] Hirst went on to compare the English situation with that in France and Italy. In these countries, he argued, ill-conceived and hasty reforms had engendered reactionary backlashes against movements to liberate geometrical teaching from the strictures of Euclid. Therefore, he urged the Association to proceed slowly and with caution towards the goal of creating a Euclidean substitute.

As a first step, Hirst recommended that the group "agree . . . upon a certain sequence in geometrical propositions, as well as upon a strict observance of certain indispensable fundamental principles." This amount of uniformity would guarantee that people marking examinations would not be constantly forced to decide whether a student was acting legitimately when using one theorem to prove another. The actual proofs would not be prescribed, however. Any demonstration which was sound would be accepted. As Hirst put it: "If we can come to this understanding, then, without interfering with freedom of exposition, we shall secure sufficient uniformity to enable examiners . . . to test the knowledge of candidates in a perfectly general and impartial manner."[14] Thus, the Association initially undertook the apparently modest task of preparing a syllabus which would establish the order in which geometrical theorems should be presented and proved. Despite the restrained nature of this goal, however, it proved difficult to attain. It was five years before the syllabus had been examined and reexamined by enough different people and groups that it was finally deemed acceptable and published in 1875.[15]

The problem was that, despite universal agreement that geometry was an absolutely unambiguous study, actually reaching a consensus about its nature, without reference to Euclidean authority, proved very difficult. Therefore, even in the early years, some members of the Association suggested that the group abandon the attempt to agree on a single syllabus and tailor geometrical teaching in a variety

---

[13] Association for the Improvement of Geometrical Teaching (AIGT), *First Annual Report*, (Birmingham: Josiah Allen, 1871), p. 9.

[14] Ibid., pp. 12–13.

[15] AIGT, *Syllabus of Plane Geometry*, (London: Macmillan and Co., 1875).

of ways to meet the needs of different students. An appendix to the second report contained the following comments by Wallis Hay Laverty, Fellow and Mathematical Lecturer at Queen's College, Oxford.

> It is submitted to Members of the Association:
>
> 1. That it is desirable that the Committee proceed at once to determine the scope of the proposed Text-book: that is,
>   A. Whether it should be one to satisfy the Metaphysicians;
>   B. Or simply an efficient Manual.
> 2. That to compromise this question will prove futile, as the result will be, at every step, to revive discussion upon method.
> 3. That no body of men short of the Metaphysicians themselves can produce a book which will satisfy the Metaphysicians; who, moreover, differ as widely as the poles upon the fundamental questions of Geometry.[16]

Laverty's suggestion that all might not agree about the essential nature of geometry, here couched as creating a rift between teachers and metaphysicians, was pursued in the general meeting in more practical terms. There J. F. Iselin, who was inspector of schools connected with the Department of Science and Art, suggested that those headed for practical careers such as architecture, surveying, carpentry or mechanics be spared the philosophical subtleties of theoretical geometry. He essentially recommended that the commonsensical geometers like Wilson not be forced to answer to the more rigid standards of someone like De Morgan.

The kinds of issues which faced English educators who were trying to teach geometry for practical ends can be seen in the teaching of constructions. Many of the reformers wanted to relax the Euclidean restrictions which limited geometrical constructions to compass and straightedge. To quote from E. M. Reynolds, who published a text entitled *Modern Methods in Elementary Geometry* in 1868:

> The arbitrary restrictions of Euclid involve him in various inconsistencies, and exclude his construction from use. When, for instance,

---

[16] AIGT, *Second Annual Report,* (Birmingham: Josiah Allen, 1872), "Appendix" p. 28.

in order to mark off a length upon a straight line, he requires us to describe five circles, an equilateral triangle, one straight line of limited, and two of unlimited length, he condemns his system to a divorce from practice at once and from sound reason. The constructions given in this book are theoretically consistent, and are employed by practical men.[17]

Certainly Reynold's wish to modify the Euclidean approach for his elementary teaching seems reasonable.

Such suggestions were not really sensible, however, within the English approach to geometry. In this view, rigorous geometry was the exemplar of clear truth and sound reasoning. These absolutes were seen to be universal, equally important for all people no matter what their career goals. Thus, Hirst resisted all suggestions that geometry might legitimately be treated in different ways to meet different needs. He pointed to Italy where, he claimed, "the subject of technical instruction had been fully considered . . . first instruction in geometry to pupils in both their technical and their classical schools, was of a strictly rigorous character."[18] He firmly maintained that the same goal could be reached in England. This conviction was ultimately embodied in the single syllabus for plane geometry which the AIGT published.

The unity of the AIGT's output was achieved at a price, however. In its early years the Association's *Reports* resound with caveats emphasizing that no single group could be totally happy with the finished product because the syllabus being prepared represented a compromise among various points of view. In the end it appears that the finished syllabus really pleased no one. For the reformers it was too conservative: to quote the disappointed Henrici, "There is very little of the influence of modern ideas to be found."[19] Bland though it was from this point of view, however, the syllabus was still too innovative to satisfy the conservative mathematical community which defended Euclid. Until the end of the century, the Euclidean,

---

[17] E. M. Reynolds, *Modern Methods in Elementary Geometry*, (London: Macmillan and Co., 1868), p. vi.

[18] AIGT, *Second Report*, (1872), p. 22.

[19] Olaus Henrici, "Presidential Address: Mathematics and Physics Section," *Report of the Fifty-third Meeting of the BAAS held at Southport in September, 1883*, (London: John Murray, 1884), pp. 393–400; reprinted in *Nature*, 28 (1883):500.

as opposed to the AIGT's, order of propositions remained central
for the mathematical examinations at Cambridge and Oxford. Since
these examinations effectively set the standards for secondary
school curricula, this refusal to change was a major defeat.
Although the AIGT continued to meet, and published annual
reports through 1893, it did not succeed in creating revolutionary
changes in geometrical education.[20]

The wishy-washy nature of the AIGT's product combined with its
institutional failure makes it tempting to pass over the group as
unimportant, the discussion over which it presided as unfruitful.
However, the issues which were brought to light in its deliberations
were often fundamental. As Hirst's statement of goals indicates, the
reformers were not merely trying to write educational pablum, readily digestible by a variety of students. They were trying to present
geometry "of a strictly rigorous character" in their reformed textbooks. The discussions surrounding these texts, then, often focused
on questions of mathematical rigor. The late-century English interpretation of this perennially slippery term can be quite finely delin-
eated by looking at the shape of the textbook debates.

GEOMETRY, SOUND REASONING AND CLEAR CONCEPTION

When the reform effort initially got underway, the common starting
point for considerations of geometry in education was the unitary
view of truth described in the first chapter. In this view, geometry
was a study closely linked to all other forms of human knowledge.
It was educationally valuable because it served as an exemplar of
real knowledge; learning it would show students the best ways to
attain that knowledge. This approach to geometry was the common
ground on which Wilson and De Morgan staged their disagreement.
It was ubiquitous, even providing the starting point for Huxley's
attack on the value of mathematical study at all.

In natural theology, the emphasis in this connection between

---

[20] W. H. Brock, "Geometry and the Universities: Euclid and his Modern Rivals, 1860–1901,"
History of Education, 4 (1975):21–35.

geometry and other forms of knowledge was on the nature of geometrical truth. This was the issue in the Stephen-Ward discussion of necessary truth, for example. The educational debates of the 1870s and 1880s focused more attention on geometry as an exemplar of reasoning. Within the context of a psychology of faculties, educators emphasized geometry's value as training in how to think and attain truth. This methodological emphasis can be seen in Huxley's attack on mathematics.

Huxley's educational philosophy was grounded in a belief that human cognition resulted from the interaction of several independent faculties or skills, like logic, observation or memory, which could be applied to a variety of subject matters. Education was a process which exercised and developed the various faculties so they would be strong and vigorous in the face of the real difficulties of life. Huxley's attack on the value of mathematical education depended on his insistence that it was merely a deductive study. This made the subject methodologically narrow and unique, thereby jeopardizing its claims to educational usefulness.

Sylvester's response was to reintegrate geometry into the intellectual picture by tempering the deductive view of its methodology. He focused on the connections between geometrical and other common forms of reasoning. Within his liberal educational ideology, emphasizing these connections meant stressing the great variety of approaches which would constitute sound geometrical reasoning.

Sylvester used projective geometry as his illustration. Though their subject matter was more elementary, the geometrical reformers were essentially pursuing the same methodological values in their texts. This is not surprising since often it was the same people pursuing these values in two different spheres: many of the vocal advocates for the study of modern geometry, including Clifford, Hirst, Henrici and Spottiswoode, were active members of the AIGT. The connection is also reflected in the titles of many of the reformers' texts, such as *Modern Methods in Elementary Geometry, Elements Adapted to Modern Methods in Geometry* or even *Exercises on Euclid in Modern Geometry*. The writers of these elementary textbooks were trying to introduce the kind of free and diversified style of geometrical argument for which Sylvester had voiced his approval before the BAAS.

The necessity of broadening and liberalizing Euclidean geometrical arguments was a major rallying cry for the reformers. In his "Preface," Wilson virulently attacked the tight restrictions within which the Greek presented his arguments. "There is no real advantage in the arrangement of propositions . . . Euclid maintains so strictly," he argued,

> for nothing can be gained by excluding *any* sound method of reasoning . . . His [Euclid's] self-imposed restrictions, geometrical and logical, have made his Geometry confused in its arrangement, and unnatural and forced in the nature of his proofs, so too the detailed syllogistic form into which he has thrown all his reasonings, is a source of obscurity to beginners, and damaging to true geometrical freedom and power.

Euclid's restrictiveness was, Wilson continued, particularly damaging to beginning students.

> There is . . . a natural and inevitable difficulty in the task of tracing data into the inferences founded on them; the geometrical facts are new: it is new to the learner to find himself reasoning consecutively at all. If then to all this novelty we add the constant analysis into syllogisms of inferences which are obvious without this analysis, and the constant reference to general axioms and general propositions, which are no clearer in the general statement than they are in the particular instance, we make the study of Geometry unnecessarily stiff, obscure, tedious and barren.

Wilson argued that elementary geometry students should be liberated from what he termed the "iron fetters" of Euclidean reasoning.[21]

Wilson's attempt to liberalize geometrical reasoning would seem to necessitate lowering standards of rigor in the subject. Certainly one can persuasively argue for the truth of a theorem in a variety of ways, but *proving* it requires adhering to very high logical standards. In late nineteenth-century England, however, this was not clear. Geometrical theorems were based on "sound reasoning" about clear conceptions. There was no fixed canon to define what

---

[21] Wilson, *Elementary Geometry*, (1868), pp. vi–vii.

was permitted. The reformers' proposals to incorporate alternative methodologies into the study were theoretically not controversial, as long as the new forms of reasoning they introduced were truly sound. The problem was that, when put to the test, there seemed to be no clear agreement about what *was* sound.

The ultimate fluidity of the traditional English view of geometrical reasoning can be seen in the response to Wilson's claim that it was rigid and inflexible. De Morgan's response to Wilson's attack was pointed. "Euclid a book of syllogistic form! We stared. . . . A look will show the difference between Euclid and syllogistic form." He then made it irritatingly clear that Wilson ought to study logic before he misused important terms like "syllogism."

Having firmly established Wilson's ignorance of logic, from his position as England's foremost logical thinker De Morgan elaborated the proper relationship between geometrical and logical reasoning. "We especially intend to separate the logician from the geometer," he wrote.

> We know that mathematicians care no more for logic than logicians for mathematics. The two eyes of exact science are mathematics and logic: the mathematical sect puts out the logical eye, the logical sect puts out the mathematical eye; each believing that it sees better with one eye than with two.[22]

De Morgan clearly felt that both logic and geometry would benefit if their practitioners knew both subjects. However, he did not believe that either subject supported the other. They were complementary as opposed to intertwined.

De Morgan's view that geometry and logic were essentially separate subjects was common in his day. He was an active participant in developments which culminated in the development of a formal view of the nature of logic. As the study of logic was becoming more formalized, however, it was also becoming increasingly removed from the study of right reasoning. Formal logic was developing as an algebra, and like algebra, was losing its anchor in conceivable reality, in this case in sound reasoning. As was the case for algebra, however, the conceptual position and strength of geometry was unaf-

---

[22] De Morgan, "Review," (July 18, 1868), p. 71.

fected by this development. Despite a profusion of algebraic notions, geometrical space remained Euclidean; similarly, in the face of rapid developments in formal logic, geometrical proofs remained reasonable.

De Morgan perhaps expressed the distinction between logical and geometrical thought more sharply than most of his contemporaries would have. As his conflict with Wilson over the term "syllogism" suggests, he had a very finely tuned ear for accurate word usage. However, a major theme of the early AIGT meetings concerned ways to develop logical thinking as a spin-off of geometrical instruction. Their views on the subject can be distilled from these discussions.

In his early presidential addresses to the AIGT, Hirst expressly considered the relationship between logic and geometrical reasoning. At the third meeting he described it as follows. "It [logic] might figuratively be described as the scaffolding by means of which the geometrical structure has been built, rather than as the first step by which the erected building is to be entered."[23] By comparing logic to the scaffolding rather than to the steel girders of the structure, Hirst emphasized the looseness of the relationship of geometry and logic. A building stands on its own when the scaffolding is removed. It can be entered and explored directly without understanding the construction process. Geometrical reasoning could be mastered without any logical knowledge.

For educational purposes, however, it was important to emphasize and explore the logical structure with which the geometrical arguments had been erected. This required pointing out the logical principles which were exemplified by geometrical proofs. To follow Hirst as he developed his metaphor:

> It has often been justly remarked that Elementary Geometry is itself the best exercise in logic for young minds. It should not be forgotten, however, that the training is incomplete until the logic is separated, mentally, from the subject matter to which it is applied. The scaffolding . . . must be removed from the erected edifice before it can be employed in building again.[24]

---

[23] AIGT, *Third Annual Report,* (Birmingham: Josiah Allen, 1873), p. 14.
[24] Ibid.

To develop his reasoning powers, then, the student was first to learn geometrical theorems and proofs; only after he had mastered them could he come to a useful understanding of the logical framework by means of which his thoughts, and the AIGT syllabus, had been constructed.

Hirst's point of view was incorporated into the form of the AIGT syllabus. This work was preceded by a "Logical Introduction," itself preceded by a caveat.

> The Association have prefaced their syllabus by a Logical Introduction, but they do not wish to imply by this that the study of Geometry ought to be preceded by a study of the logical interdependence of associated theorems. They think that at first all the steps by which any theorem is demonstrated should be carefully gone through by the student, rather than that its truth should be inferred from the logical rules here laid down. At the same time they strongly recommend an early application of general logical principles.[25]

The general thrust of this typically ambiguous statement was to subordinate logical principles to geometrical understanding. Understanding was to be achieved in whatever way it could be; recognizing logical principles implicit in its objects came later.

From a theoretical standpoint, then, it seems that Wilson may have overstated his case when he lamented that students were being forced to reason "in iron fetters." Most of the discussants agreed that there really was no fixed form in which geometrical arguments had to be cast in order to be valid. Practically, however, his point was well taken. When actually put to the test, the mathematical community applied a devastatingly high standard of criticism to the new geometry texts. They subjected innovation to minute scrutiny, ferreting out inconsistencies, hidden assumptions and slips between steps in proofs.[26] At the same time, theirs was a double standard. They played down the fact, amply demonstrated by De Morgan himself, that Euclid's own reasoning was often flawed. Faced with a sit-

---

[25] AIGT, *Syllabus*, (1875), p. 3.
[26] As an example of the kind of care taken over the new texts, see the "Review of J. R. Morell, *The Essentials of Geometry, Plane and Solid, as taught in the French and German Schools, with shorter Demonstrations than in Euclid,*" *Nature*, 3 (1871):323–25.

uation in which no one could claim perfection, the English clung to the strictures of Euclid.

Theoretically, a single standard for geometrical reasoning was not necessary, however. Ultimately the criteria for acceptability were vague, the fetters were more like velvet than iron, and any form of argument, provided it was "sound," was acceptable. With such an ultimately unclear vision of the steps which led from one geometrical theorem to another, it is tempting to ask where the foundations of the subject rested. Textbooks, after all, were to be rigorous. The answer to this question which emerges from the textbook debates ultimately rested on conceptual clarity. However, the adequacy of this criteria of mathematical validity, which had been so freely assumed in abstruse philosophical writings, was put to a strenuous practical test in the attempts to actually produce generally acceptable new geometry textbooks.

One of the central challenges facing the writers of new textbooks was the Euclidean theory of parallels. The same kinds of obscurity in Euclid's theory which had led non-Euclidean geometers to propose their alternative geometrical systems, led educators to try to ground Euclid's theory more clearly. Their texts were often marked by alternative presentations of the classical theory of parallels, and these alternatives provided a focus for much of the textbook discussions. Again the Wilson-De Morgan exchange sets the tone for much of what followed.

One of the earliest and most radical attempts to create an alternative theory of parallels which would be more palatable for the schoolboy than Euclid's was Wilson's. In the first edition of his *Elementary Geometry*, Wilson abandoned the definitions and axioms on which Euclid had based his theory in favor of a totally new approach. The linchpin of Wilson's innovation was the basically undefinable concept of "direction." He introduced this notion in his fourth definition: "A *straight line* is a line which has the same direction at all parts of its length." It was elaborated somewhat further in the sixth definition: "Two straight lines that meet one another have different directions, and the difference of their direction is the *angle* between them."[27] From these beginnings Wilson constructed the following definition of parallels.

---

[27] Wilson, *Elementary Geometry*, (1868), p. 2.

Def. 15. Straight lines which have the *same direction,* but are not parts of the same straight line, are called *parallel lines.*

It is obvious from the definition:

1) That pairs of parallel lines are in the same plane.

2) That parallel lines would never meet however far they were produced; since if they met they would have different directions.

3) That straight lines which are parallel to the same straight line are parallel to one another.[28]

This definition, which was all entwined with the notion of direction, formed the basis for Wilson's subsequent treatment.

Wilson presented his direction-based theory of parallels as a rigorous alternative to Euclid's. "I contend," he wrote in his preface, "that it is possible to present the science of Geometry in a more natural and simple order, and demonstrate the properties of figures by proofs differing in many cases from Euclid's, without being less rigorous."[29] In making this statement, Wilson was explicit about the view of rigor he espoused. "The merit of Euclid's treatment of Geometry," he wrote, "consists in his resolutely fixing the mind on the *things* with which the argument is concerned, and never allowing the substitution of mere *symbols.*"[30] As it was for the majority of his compatriots, geometry for Wilson was descriptive; it could properly be labeled an exact science. Its rigor lay in the exact fit of the geometrical structure with the conceptual subject matter being described. He felt free to propose alternatives to Euclid's structure as long as he remained conceptually clear.

Wilson was not alone in regarding his book as a rigorous alternative to Euclid which could therefore be subjected to minute critical review. It was on these grounds that Augustus De Morgan reviewed the work in the *Athenaeum* and pronounced it a failure. The spirit of De Morgan's criticisms can best be seen by quoting at some length from his treatment of Wilson's notion of "direction."

There is in it [Wilson's *Elementary Geometry*] one great point, which brings down all the rest of it fall, and may perhaps . . . support the

---

[28] Ibid., p. 12.

[29] Ibid., pp. ix–x.

[30] Ibid., p. ix.

rest if it stand. That point is the treatment of the angle, which amounts to this, that certain notions about *direction*, taken as self-evident, are permitted to make all about angles, parallels and all, immediate consequences. . . .

What 'direction' is we are not told, except that 'straight lines which meet have different directions.' Is a direction a magnitude? Is one direction greater than another? We should suppose so; for an angle, a magnitude, a thing which is to be halved and quartered, is the 'difference of the direction' of 'two straight lines that meet one another.' A better definition follows; the 'quantity of turning' by which we pass from one direction to another. But hardly any use is made of this, and none at the commencement. And why two definitions? Is the difference of two directions the *same thing* as the rotation by which we pass from one to the other? Is the difference of position of London and Rugby a number of miles on the railroad? Yes, in a loosely-derived and popular and slipslop sense: and in like manner we say that one man *is* a pigeon pie, and another *is* a shoulder of lamb, when we describe their contributions to a picnic. But *non est geometria!* Metaphor and paronomasia can draw the car of poetry; but they tumble the waggon [sic] of geometry into the ditch.[31]

De Morgan's discussion of Wilson's attempt reflects the same descriptive view of rigor Wilson espoused. Thus it is the conceptual adequacy of Wilson's definition of direction which is the focus of De Morgan's critique. It is because Wilson failed to create an exact definition—that is, one which exactly pinpointed a spatial concept—and instead resorted to vague indications of his meaning, that De Morgan took him to task.

The Wilson-De Morgan interchange points to an important problem with the descriptive view of geometry which persistently plagued those trying to create alternatives to Euclid's text. Any innovation was open to attack for being unclear. Euclid could weather such criticisms by the sheer weight of his authority. Anyone challenging the clarity of Euclidean conceptions was going against a dauntingly long historical tradition. Innovators, on the other hand, were highly susceptible to conceptual challenges.

This is because at bottom, the conceptual criterion of rigor was

---

[31] De Morgan, "Review," (July 18, 1868), p. 72.

psychological. To attain rigor in this way, one had to pinpoint the basic, clear concepts and somehow establish subjectively which were primary and which derivative. The complications inherent in this approach can be illustrated by a passage from Charles Dodgson's play *Euclid and his Modern Rivals,* a deservedly obscure 225-page drama the Oxford don devoted to detailed critiques of contemporary textbooks. In this passage, Niemand is a purportedly neutral character who is presenting Wilson's text to the scrutiny of the commited Euclidean Minos.

Min. . . . I wish to ask you one question . . . It is, in fact, *the* crucial test as to whether 'direction' is, or is not, a logical method of proving the properties of Parallels. . . . Does the phrase 'the same direction,' when used of two Lines not known to have a common point, convey to your mind a clear geometrical conception?

Nie. Yes, we can form a clear idea of it, though we cannot define it.

Min. And is that idea (this is the crucial question) *independent of all subsequent knowledge of the properties of Parallels?*

Nie. We believe so.

Min. . . . You feel certain you are not unconsciously picturing the Lines to yourself as being equidistant, for instance?

Nie. No, they suggest no such idea to us.

Min. . . . But do you feel equally certain that you are not unconsciously using your subsequent knowledge that Lines exist which make equal angles with all transversals?

Nie. We are not so clear about *that.* It is, of course, extremely difficult to divest one's mind of all later knowledge, and to place oneself in the mental attitude of one who is totally ignorant of the subject.

Min. Very difficult, no doubt, but absolutely essential, if you mean to write a book adapted to the use of beginners. My own belief as to the course of thought needed to grasp the theory of 'direction' is this:—first you grasp the idea of 'the same direction' as regards Lines which have a common point; next, you convince yourself, by some *other* means, that different Lines exist which make equal angles with all transversals; thirdly, you go back, armed with this new piece of knowledge, and use it unconsciously, in forming an idea of 'the same direction' as regards different Lines.[32]

---

[32] Dodgson, *Euclid,* (1885), pp. 129–131.

It is certainly tempting to agree with Niemand about the extreme difficulty of "divesting one's mind of all later knowledge" in order to judge the conceptual purity of a geometrical definition. His difficulties along these lines are strongly reminiscent of those with which Cayley was struggling when he tried to evaluate the concept of distance in his "Note."

Not only was this kind of argument about conceptual clarity inherently too slippery for response, it was by nature highly conservative. It proved virtually impossible to counter the claims to clear conception of someone who rested on a work as venerable as Euclid's *Elements* and failed to see any alternative approach as clearly. Dodgson's book is full of instances where he adjudges less radical innovations than Wilson's to be adequate but still unnecessary, since he feels Euclid had done as well. In fact, at the beginning of his play Dodgson makes this resistance to change an explicit part of his program. There he has Euclid say, "Unless, then, it should appear that one of my Modern Rivals, whose logical sequence is incompatible with mine, is so decidedly better in his treatment of really important topics, as to make it worth while to suffer all the inconvenience of a change of numbers, you would not recognise his demand to supersede my Manual?"[33] Given these difficulties, it is not hard to understand why in their syllabus, the AIGT decided to treat parallels "in Euclid's manner" and why Wilson edited the idea of "direction" out of the third and subsequent editions of his book.

The immediate outcome of this disagreement, then, was capitulation on the part of the innovators and a forced retreat to the conceptual clarity guaranteed by the authority of Euclid. Nevertheless, the experience of the controversy suggested the arbitrariness of that position and pointed to the weakness of the conceptual view of rigor on which it was grounded. Geometry had for decades been touted as the exemplar of perfect and indisputable truth. The very fact that England's foremost mathematicians could fiercely and persistently disagree about the essential validity of different elementary presentations of the subject raised questions about this time-honored assumption. The innovators lost the battle on the treatment of parallels, as they did on many other issues, but their dissatisfaction with

---

[33] Ibid., p. 12.

this retreat prepared the way for radically new criteria for geometrical validity which were introduced at the turn of the century.

EDWARD TRAVERS DIXON AND THE FOUNDATIONS OF GEOMETRY

The intellectual dynamic created by the apparently mundane textbook discussions of the 1870s and 1880s can be illustrated in the work of Edward Travers Dixon. In 1891, Dixon, who was a student at Cambridge at the time, published a book entitled *The Foundations of Geometry*. In it he proposed a nonconceptual view of the nature of geometry as a way to circumvent the dead-end discussions about clear conception which had bogged down educational innovators like Wilson. Dixon followed this book with *An Essay on Reasoning*, also published in 1891, in which he tried to further explain and elaborate the ideas of his *Foundations*. In the next two years, he continued to defend his views in public letters and articles.

Dixon was ultimately unable to meet the challenges inherent in the new view of geometrical rigor which was implicit in his work. He was effectively silenced by the objections which were raised in the rather desultory discussion of his ideas. In many ways his views were flawed, and he was not able to carefully and consistently develop them. In short, Dixon should not be rediscovered as a neglected seminal figure of late nineteenth-century mathematics.

In the attempt to see the directions in which English educational conflicts were leading, however, Dixon's very weakness is his strength. He can serve as a weather vane pointing in the direction from which the winds of change were blowing. The direction in which he points is remarkably similar to the direction in which retrospective observers have asserted non-Euclidean and projective developments were moving.

Dixon's *Foundations of Geometry* was intended as a clarification of the nature of geometry for the use of high school teachers. In the unitary spirit of the British educational tradition, however, its intended audience did not temper Dixon's basic goal, which was to rewrite Euclid's geometry in order to place it on a rigorously sound foundation. To quote the opening words of the "Preface": "I believe that the system of geometry I have set forth in this book is logically sound, and that consequently the more it is discussed and criticised, the more firmly will it become established."[34]

Dixon's work is divided into three parts which are directed towards three somewhat different ends. The first aims to clarify once and for all the nature of geometrical truth. The second contains what Dixon hoped was the definitive definitional and axiomatic structure of basic geometry. The third is an attempt to assess the applications of his theoretical structure for real space.

In this book, Dixon betrays very little knowledge of the geometrical developments being pursued at the forefront of research. In his third section he focuses on four-dimensional geometry and its implications for the physical study of space, but his understanding of these developments seems basically elementary. He does not seriously consider a non-Euclidean theory of parallels; in fact, he proves the parallel postulate in his second section. Nor does he evince any knowledge of or interest in projective geometry. Nonetheless the book's central argument focuses on the same philosophical questions concerning the nature of geometry which were being raised by these new developments.

It is clear from his opening words that Dixon's interest in the foundations of geometry stems from an interest in the more general epistemological question of how humans can know truth.

> Can we be absolutely certain of anything in this world? Or is all our knowledge only empirical and approximate? Is there such a thing as *necessary truth,* and if so how are we to know when we have attained it?
>
> These questions open up perhaps the most disputed branches of Logic and Metaphysics. Under one form or another the contest has been raging round them ever since the time of Aristotle. The line of battle has sometimes shifted forward, sometimes back, sometimes it has changed front, so that quite new issues seem to be at stake. But the status of the Science of Geometry has always been the key of the position; though the combatants on both sides have often confined

---

[34] Edward T[ravers] Dixon, *The Foundations of Geometry,* (Cambridge: Deighton, Bell and Co., 1891), p. iii.

their energies to flank attacks, in despair of making any impression on the citadel.[35]

Dixon's first part, entitled "On the Logical Status of Geometry," is essentially an attack on the citadel of necessary truth based on a reevaluation of the nature of geometrical knowledge.

Dixon's basic strategy was to undercut the whole notion of necessary truth by drawing a sharp distinction between subjective and objective knowledge. Within the epistemological tradition stretching back to Locke, the special function of geometrical knowledge and of the truth it exemplified was as a bridge between these seemingly irrevocably separated realms. Dixon attempted to destroy this bridge by arguing that all things which could be conceived clearly enough to meet the criteria for necessary truth were, by their nature, confined to the subjective sphere.

> Each of us can be absolutely certain of what he is feeling or conceiving at any given moment, and if there are any of his conceptions which he can call up at will he can be absolutely certain that those conceptions have to him a real existence; subjectively, in his own mind, be it understood; though not necessarily objectively, outside it.[36]

From this point of view, then, insofar as it was necessarily true, geometry was merely a subjective system. Any geometrical statements about objective, physical space, however, were merely contingent. In the third section of his book, where he applied the geometrical system of the second to physical space, Dixon asserted that the objective truth of geometry "is proved by induction as convincing as any we know of, except perhaps that which convinces us that there is an objective universe at all."[37] Sure though this knowledge may be, however, it falls short of necessary truth. Thus, ultimately, Dixon argued, the objective truth of geometry is of the same character as that to be found in any of the inductive sciences. With this

---

[35] Ibid., p. 1.
[36] Ibid., p. 3.
[37] Ibid., p. 143.

point, Dixon was defending a basic empirical position similar to that asserted by J. S. Mill.

Ultimately, however, Mill's view of geometry was descriptive. As was emphasized in Chapter 1, one of Mill's rare points of agreement with his idealistic adversaries was that geometry was the study of a conceptual entity which existed independently of mathematical developments. Among other things, his assertion that geometrical theorems described conceptual space allowed geometry to keep its status as an important part of an integrated view of knowledge.

Dixon, however, raised serious issues for this descriptive view. His originality emerges in the second section of his book, where he tried to construct an actual geometry. In this geometrical presentation, Dixon resurrected Wilson's idea that a notion of "direction" was the proper foundation for the theory of parallels. In order to avoid being trapped in the kind of discussion which had so devastated Wilson, however, Dixon devoted much of his first philosophical section to recasting the notion of what a geometry entailed. In particular, he tried to change the criteria for what constituted an adequate geometrical definition.

Dixon's discussion of definitions flows from his initial argument that all subjects which concerned necessary truth were wholly subjective. This led him to discuss the foundations of such sciences. A subjective science, he explained, was one in which the premises were directly apprehended by the mind. These premises could be classified under four headings: objective facts, postulates, axioms and definitions.

Following the lead of many philosophical predecessors, Dixon turned his attention to definitions as the key to understanding the nature of geometry. He specified three important roles which a definition could fill. As the foundations of a subjective science, Dixon insisted that definitions were not merely verbal or tautological. In the first place, he claimed, knowledge could actually be increased by the introduction of a careful definition. Thus, a clear definition of "even number," where no such concept had been recognized before, would open whole new arithmetic vistas theretofore unexplored. Secondly and relatedly, he maintained that the creation of clear definitions could serve to clarify and specify concepts only vaguely understood before.

These two uses of definitions fit easily into traditional views of geometry as a conceptual study. The third point Dixon emphasized about definitions leads in a radically new direction, however. Whereas in the first two points he emphasized ways definitions might introduce or clarify concepts, in his third point Dixon asserted that definitions could also be formulated in such a way that they transcended concepts altogether. "But lastly," he continued,

> it is a characteristic of that marvelous instrument, Language, that by its aid we are actually able to reason accurately about things we do not clearly conceive; and this we are enabled to do by defining a word or symbol *by its attributes,* without necessarily conceiving its meaning as a whole at all.[38]

Using this kind of definition, which Dixon called an implicit definition, one can reason about things without asserting either their objective or their subjective reality.

> Thus all that is logically required for a definition is one or more assertions with regard to the word to be defined or its attributes. Having made the assertions it may be that there is no objective reality corresponding to the word. . . . It may further happen that we know of no subjective concept corresponding to it; . . . but unless the assertions are inconsistent, that is unless the falsehood of one can be deduced from the truth of another, the definition remains logically sound, and we may deduce theorems from it which may, or may not, turn out useful or interesting, but which at least are logically true.[39]

Dixon's emphasis on "logical truth" in a geometrical setting was a radical departure from previous geometrical practice. Recognizing this kind of definition as legitimate in geometry would shatter the descriptive view which had supported the study for so long. He was suggesting that geometry could be pursued definitionally, that its definitions could truly create its objects rather than merely describing them.

The impetus behind this radical suggestion of Dixon's was a deter-

---

[38] Ibid., p. 5.
[39] Ibid., pp. 5–6.

mination to move the discussions of elementary geometry away from the bog of conceptual argument and onto the unambiguously firm ground of logical adequacy. He hoped that establishing the validity of this third kind of definition would allow him to avoid the confusions Wilson had encountered as he tried to defend his definition of direction. Dixon preceded his geometrical section with a "Note" asking specifically that the arguments against Wilson not be "dished up" again.[40] In his "Preface," Dixon went even further in his attempt to control the response to his work. There he laid out the basic issues he felt discussants could legitimately address in assessing its value.

(i) Do you accept the requirements I have laid down for a logical definition? . . . (If not, please state which of them you object to, why you object to it, and what you propose to substitute for it.)

(ii) Do you entertain a mental concept (which I shall call by the name 'direction') such that the assertion "A Vector is a given amount of transference in a given direction, irrespective of the point of departure," is intelligible to you?

(iii) If so, does not this concept fulfill all the four requirements of my definition of 'direction'?[41]

Thus, Dixon's views of definition were developed in an attempt to constructively shape the discussion he was hoping his work would generate.

Despite his careful preparation for it, Dixon's book did not occasion a great deal of response. Though he was applauded after he presented his ideas to the AIGT in 1891, the chairman described the discussion as "desultory."[42] What comments were made—at the AIGT meeting or in the review in *Nature*—did not follow his guidelines and did resurrect the kinds of issues which had been occasioned by Wilson's remarks twenty years before. This can be seen, for example, in the response to Dixon's implicit definition, "A direction may be conceived to be indicated by naming two points, as the direction from one to the other." The reviewer for *Nature*

[40] Ibid., p. 31.

[41] Ibid., p. iii.

[42] AIGT, *Seventeenth General Report*, (Bedford: W. J. Robinson, 1891), p. 42.

dismissed this definition with the remark that "after repeated mental effort, we are still unable to realize the meaning of the 'direction from A to B' without using the concept of the 'straight line AB.'"[43] Despite his efforts to lift the focus of the discussion out of the conceptual sphere, Dixon's educational respondents kept it there.

The fault for the persistent conceptual emphasis does not lie only with the educators, however. The formal thrust of Dixon's argument itself was consistently muddied by a residual conceptual overlay. So, for example, in the second of the steps for critics quoted above, Dixon has already returned to conceptual criteria to ground geometry. Similarly in a passage in the text, he explains that definitions of terms like "direction" would have to first be logically consistent and independent but also be true to the term's original meaning. To quote his rather convoluted terms:

> Where it is required to define logically a term whose denotation is already known, it is further necessary not only that (iii) the assertions should be commonly accepted as true with respect to it, but that (iv) they should restrict the meaning of the term exactly to its accepted denotation, neither more nor less, and should do so in the simplest manner that can be devised.[44]

With this prevarication, Dixon clearly opened the door for just the kind of conceptual discussion his formal view of definitions might have allowed him to avoid.

In the writings which followed *The Foundations of Geometry*, Dixon continued to insist on the importance of formal definitions, though he never rid himself of his residual conceptual baggage. Just months after he published *The Foundations*, he followed it with *An Essay on Reasoning* in which he tried to be clearer about the kind of argument he felt formed the backbone of geometry. In his chapter entitled "Definitions," Dixon argued even more explicitly that to view definitions as precise indicators of a concept being defined was an error.

---

[43] E.M.L. [probably E. M. Langley, the mathematical master of the Modern School in Bedford and an AIGT activist], "Review of Dixon's *The Foundations of Geometry*," *Nature*, 43 (1891):554.

[44] Dixon, *Foundations*, p. 21.

"The process of suggesting to the mind the [conceptual] import of terms is called *description* not definition," he insisted. Definitions did not serve a descriptive function. They were valuable only insofar as they enabled the practitioner to clearly ascertain what did or did not fit their terms.

> The object of a logical definition is only to provide a test by which we may determine whether a given thing or idea is, or is not, included in the meaning of the term. It is not even necessary in a symbolic argument that all the terms should have import [conceptual meaning] at all, as long as it cannot be formally shown that they are incapable of having any, by reason of their definitions involving contradictions in terms.[45]

This kind of argument was not new in the context of algebra where, at least practically, many had moved beyond arguing over the conceptual meaning of each symbol use. But Dixon was trying to apply this analysis to geometry and to geometrical reasoning, which had always been descriptive. In this, Dixon was pushing beyond the notion that geometry was merely subjective to the notion that it did not necessarily have any conceptual or objective import at all.

This point drew the fire of another of Dixon's reviewers, who insisted in *Mind* on the importance of the concept to the initial creation of the list of attributes which formed the implicit definition. "But," the reviewer objected,

> how can we legitimately make these assertions without first assigning a *meaning*, i.e., at least *subjective import* . . . to the term? After having tested the truth of the assertions, . . . it is true that we can temporarily lay aside all thought of the meaning, and proceed to make logical deductions. But the arbitrariness of the assertions applies only to the choice of the term or symbol . . . That the assertions are true of the concept is not arbitrary.[46]

---

[45] Edward T[ravers] Dixon, *An Essay on Reasoning*, (Cambridge: Deighton, Bell and Co., 1891), p. 53.
[46] *Mind*, n.s. 1 (1892):286.

A similar point was raised by E. E. Constance Jones in a correspondence with Dixon carried on in the pages of *Nature*.

> And when Mr. Dixon says that the proposition 'an isosceles triangle has two equal sides' has 'wide applicability and usefulness' because we 'often find things which can fairly be called isosceles triangles,' it seems clear that he himself cannot have taken the proposition as starting in a sense purely 'symbolic' . . . If he did, it would be little less than miraculous that an entirely arbitrary definition should happen so to fit actual experience.[47]

Clearly both of these respondents demanded that the conceptual genesis and meaning of geometrical terms continue to be pivotal in establishing the subject.

Dixon, however, was undaunted by their objections. To Jones he replied:

> I maintain that definitions should be *arbitrary,* but not necessarily that they should be made *at random* . . . The definitions of geometry could not now be laid down at random, but they are none the less arbitrary, for they require no support from any *a priori* considerations.[48]

It appears from this passage that Dixon was willing to consider geometry as a purely formal study. He seemed to be trying to move the focus away from geometry's conceptual truth and place it on the definitions themselves. It was, however, not an easy change to make.

Dixon made his most mature statement affirming the essentially formal nature of mathematical reasoning in his article "On the Distinction between Real and Verbal Propositions." Here he boldly asserted that

> the old classifications of propositions and sciences must be modified. The old division of propositions was into analytic and synthetic, of

---

[47] E. E. Constance Jones, "Letter," *Nature*, 47 (1892):78.
[48] E. T. Dixon, "Letter," *Nature*, 47 (1892):127.

knowledge into *a priori* and *a posteriori*. But we have seen that the primary distinction . . . is that between real, and arbitrary or verbal propositions; and sciences must therefore be classified according as their conclusions belong to the one category or to the other.[49]

Mathematical sciences, Dixon insisted, were purely verbal—essentially requiring neither subjective nor objective referents as the basis of their truth. His purpose in this article was to emphasize this point—to clarify "the essential distinction between both these kinds of knowledge [objective and subjective] and purely formal conclusions such as those of mathematics."[50] Unfortunately for the strength of Dixon's position, however, it is at best unclear in this paper that he would include geometry as a mathematical science. It seems that he might reserve this category for symbolic algebras and logic.

Ultimately Dixon's work was too fluid and contradictory to generate prolonged discussion or to initiate significant changes in English views of geometry. His ideas definitely seem to have been leading away from the descriptive view of geometry, however, and what response his work did generate was directed towards this issue. A similarly formal view was also brought to the AIGT's attention two years after Dixon's report to them by Gino Loria, who was then professor of higher geometry at the University of Genoa.

Loria's ideas were couched in the form of a review of the AIGT syllabus which had been prepared on projective geometry. He brought to bear a continental view of the subject and was highly critical of the AIGT's syllabus. The treatment of imaginary points and points at infinity was particularly inadequate from his perspective. He referred to a whole body of continental works by leading European geometers like Pasch, von Staudt, Segré and Lüroth which were directed towards creating a formally consistent view of all of the elements of projective space. The heart of their achievement lay in ridding projective geometry of any trace of the inductive principle of continuity which still lay at the heart of many English

---

[49] E. T. Dixon, "On the Distinction between Real and Verbal Propositions," *Mind*, n.s. 2 (1893):345.

[50] Ibid., p. 346.

treatments. In his closing sentence, Loria firmly emphasized this point:

> *The aforesaid representations of the elements at infinity with the imaginary ones render the principle of continuity superfluous,* a principle which though for the past and the present excellent as a method of discovery, cannot in any way pretend to serve as an ingredient to a strict treatment of geometry, the attainment of which is one of the noblest aims of modern studies.[51]

Loria's statement highlights once again the tension English educational emphases generated for formal, definitional views of geometry. His priorities in establishing the nature of a "strict treatment of geometry" were very different from those of a classroom teacher. Teaching the process of discovery was precisely the goal of most English geometry teachers. A reformulation of projective geometry which would eliminate the wonderfully fecund principle of continuity was not an attractive option.

In the event, Loria's review appeared in the last of the AIGT *Reports* and appears to have attracted no comment. It had already been noted in the previous meeting of 1892 that the syllabus-writing committees were making "very little progress" and that "when the present Committees have each finished the Syllabus on which it is engaged, no new ones should be appointed."[52] The group ultimately failed to achieve the goal so firmly propounded when it first was established, which was to write a single syllabus for geometry which would be sufficiently solid to please all interested parties. Dixon's and Loria's ideas might have achieved the goal from one perspective: their more formal view of the subject at least carried the seeds for the growth of more constructive discussions with objective rather than subjective criteria. It was a highly unattractive approach for educators, however, because it raised the specter of mathematical irrelevance which Huxley's speeches had invoked more than a quarter of a century before.

When curricular reform was finally introduced in 1902, it came

---

[51] Gino Loria, "A Few Remarks on the 'Syllabus of Modern Plane Geometry'," AIGT, *Nineteenth General Report*, (Bedford: W. J. Robinson, 1893), p. 53.

[52] AIGT, *Eighteenth General Report*, (Bedford: W. J. Robinson, 1892):5.

from a different quarter entirely. Euclidean education was finally routed by a renewed movement for geometrical reform led by John Perry, who was, at the time, teacher at Finsbury Technical College. The view of geometry Perry propounded was diametrically opposed to the formal one indicated by Dixon and Loria. His success was largely predicated on abandoning the unitary view of knowledge and thereby changing the whole basis of the geometrical discussion.

JOHN PERRY AND THE BREAKDOWN OF THE UNITARY VIEW OF TRUTH

The demise of the AIGT *Reports* signaled the end to efforts directed at writing geometrical syllabi. There remained, however, a strong interest in discussing geometry from an educational perspective. This interest was embodied in the *Mathematical Gazette,* which first appeared in April 1894, edited by long-time AIGT member and honorary secretary E. M. Langley. The goal of this journal was to establish communication among teachers about effective teaching methods and approaches.

At its inception, the *Gazette* was formally connected with the AIGT, which continued to meet throughout the century. In 1897, the name of the group was changed to the Mathematical Association, which signaled the expansion of its interest beyond geometry and the vexed question of Euclidean teaching. The specific issue of geometrical teaching was not completely dead, however. Discussion continued, and in 1901, interest was again fanned into flame with a speech John Perry delivered to the BAAS in Glasgow.

In this speech, Perry advocated abandoning the unitary view of truth which had been the basis for virtually all previous discussions of the nature of geometry. The impetus for Perry's attack on the unitary view of truth was mounting frustration with the restraints placed on the teaching of practical geometry by an examination system which was defined by the concerns of pure mathematicians. Jaded by the failure of more that twenty-five years of discussion, Perry did not believe that there was any possibility of reconciling the two groups' points of view. He willingly abandoned the program

Hirst had built upon his conviction that there was a single way to pursue geometry. "What we want," Perry thundered,

> is a great Toleration Act which will allow us all to pursue our own ideals, taking each from the other what he can in the way of mental help. We do not want to interfere with the students of pure mathematics, men whose peculiar mental processes are suited to these studies. . . . The more they hold themselves in their studies as a race of demigods apart the better it may be for the world. . . .
>
> I belong to a great body of men who apply the principles of mathematics in physical science and engineering; I belong to the very much greater body of men who may be called persons of average intelligence. In each of these capacities I need mental training and also mathematical knowledge.[53]

Clearly Perry spoke out of prolonged frustration at having to fit his geometrical teaching at Finsbury Technical College into the rarefied forms of argument required on standardized examinations. His reaction was not merely idiosyncratic. His talk generated a three-hour discussion of which one observer, R. F. Muirheed, noted:

> there could be no question as to the sustained interest of the debate both on the part of the speakers and on that of the listeners, who must have numbered about two hundred. . . . It was only by restricting the time of each debater to ten minutes that the Chairman was able to call on all those who had in advance promised to take part in the discussion; and one could feel that a large portion of the silent members were so perforce, and would dearly have liked to unburthen their minds of those thoughts and feelings that were stirred up in the course of the debate.[54]

One of the major themes running through the debate was the oppressiveness of educational structures which forced a single authority on all students and educators. Thus Muirheed closed his brief report with the opinion that "emancipating the teacher from

---

[53] BAAS, *Discussion on the teaching of Mathematics which took place in Glasgow, 1901*, John Perry, ed. (London: Macmillan and Co., 1902), pp. 3–4.
[54] R. F. Muirheed, "The Teaching of Mathematics," *Mathematical Gazette*, 2 (1901):82.

all detailed syllabuses, is a more hopeful [line of reform than] merely laying down a fresh syllabus, which in the nature of things will in course of time be as much of a clog upon educational progress as that which it replaces."[55]

The practical effect of this discussion was virtually immediate. Again committees were formed in both the BAAS and the Mathematical Association to put pressure on the universities to modify their Euclidean-based examinations. In 1903 Cambridge agreed to accept "any proof of the proposition, which appears to the Examiners to form part of a systematic treatment of the subject." The other major universities quickly followed suit. The hegemony of Euclidean geometry in English education came to an abrupt end[56] because the group for whom geometry was part of a practical education finally broke the power of those who defined its value strictly in terms of the liberal education.

This development allowed a move away from the mid-century view which integrated geometry into a carefully woven tapestry of human knowledge. By allowing mathematicians and educators to treat geometry differently, Perry essentially cut the rope which had joined men like Wilson and De Morgan in their ideological tug of war. He dissolved their common assumption that geometry was valuable insofar as it exemplified legitimate reasoning and knowledge.

In this, Perry's aim was to free educators from the tight restrictions of the pure mathematicians. His "toleration act" was designed to protect those who were interested in practical mathematics from the esoteric demands of rigor. The split he effected was equally liberating for the mathematicians, however. Dixon's and Loria's ideas, which sharply differentiated mathematics from other more descriptive subjects, are indications of the directions in which mathematical thought was moving. The kind of formal approach they were suggesting was clearly inappropriate for the traditional setting of the liberal education. It was highly attractive from a different perspective, however. The philosophical and social development of this kind of mathematics will form the subject of the next two chapters.

---

[55] Ibid., p. 83.
[56] Brock, "Geometry," (1975), pp. 30–31.

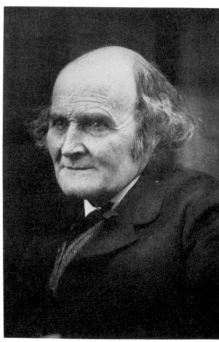

Bertrand Russell. *(Courtesy of the B. B. C. Hulton Picture Library.)*

Arthur Cayley. *(Courtesy of the Master and Fellows of Trinity College Cambridge.)*

CHAPTER 5

Bertrand Russell and the Cambridge Tradition

When viewed from the perspective of the AIGT, the rigidity of the English examination system was harmful because it imposed unreasonably high demands on the forms of geometrical reasoning required from elementary students. In the final decades of the century, England's most advanced mathematical students also chafed under the rigidities of that system, though. Their grievances came from almost the opposite direction. Whereas the members of the AIGT complained about the strictures of the matriculation examinations, the mathematical group directed their criticisms against the Mathematical Tripos. Whereas the elementary teachers were concerned that examinations were unreasonably stringent, the bright mathematics students often complained that, by contemporary, continental standards, the material on the Tripos was treated too loosely. The way geometry was tested in the late nineteenth-century English examination system was as confining for those who wanted to develop mathematics at its highest levels as it was for those trying to teach it to beginners.

This side of the educational picture was recognized by many of those who were pushing for reform. Even John Perry, in a mellow moment, noted that "the greatest sufferer hitherto . . . has been the real mathematician, who is drilled so long on elementary work that even after he becomes a wrangler [a winner of the highest honors in mathematics at Cambridge] he is only ready to begin that higher

work which he might have studied years before.''[1] Judging from retrospective reports, this seems to be an accurate representation of
the situation for the mathematically inclined. Dissatisfaction with
the mathematics which appeared on the Tripos is an almost universal theme in the memoirs of those who were educated late in the
nineteenth century.

During this period, the examination continued to be divided into
two parts, the first of which was elementary, the second advanced.
The first part retained its nature as a vehicle for a liberal education
and was therefore highly geometrical. Euclid's *Elements* and Newton's geometrically constructed *Principia* continued to form its core.

This geometrical emphasis meant that the kind of mathematical
work required to do well on the examination was not the same as
that which would be used in other contexts. To quote from a frus-
trated graduate looking back on the education of the period:

> In patriotic duty bound, the Cambridge of Newton adhered to New-
> ton's fluxions, to Newton's geometry, to the very text of Newton's
> *Principia*: in my own Tripos of 1881 we were expected to know any
> lemma in that great work by its number alone, as if it were one of the
> commandments or the 100th Psalm. Thus English mathematics were
> isolated: Cambridge became a school that was self-satisfied, self-sup-
> porting, self-content, almost marooned in its limitations.[2]

Whereas the author of this quotation, A. R. Forsyth, focused his
complaint on the isolation and irrelevance of the Cambridge education, Bertrand Russell, who took the first part of the Tripos in
1892, complained about the inadequacy of the mathematics itself.
Looking back on the experience more than sixty years later, Russell
recalled:

> The mathematical teaching at Cambridge when I was an undergrad-
> uate was definitely bad. . . . The necessity for nice discrimination
> between the abilities of different examinees led to an emphasis on
> 'problems' [memorized] as opposed to 'bookwork' [freeform]. The
> 'proofs' that were offered of mathematical theorems were an insult
> to the logical intelligence. Indeed, the whole subject of mathematics

---

[1] John Perry, "The Mathematical Tripos at Cambridge," *Nature*, 75(1907):274.
[2] A. R. Forsyth, "Old Tripos Days at Cambridge," *The Mathematical Gazette*, 19 (1935):167.

was presented as a series of clever tricks by which to pile up marks in the Tripos.[3]

It seems clear from these and innumerable similar complaints that, by the final decades of the century, the once venerable Tripos had become grotesque, its character as a contorted and outdated obstacle far overshadowing its educational function.[4]

Even while acknowledging the legitimacy of these complaints, however, it is important to recognize the theoretical underpinnings of the Cambridge Tripos and their effects on those who were involved in the system over which it loomed. On an important level, the examination continued to be an institutional manifestation of the view that mathematical reasoning, especially geometry, encapsulated the essence of sound thinking. This is the other side of Russell's complaint about the mathematical proofs being an insult to the logical intelligence. The study was pursued to encourage sound reasoning, not logical thinking. The English tradition which interpreted mathematics as an integral part of the liberal education continued to be very powerful until the early twentieth century.

Despite his disclaimers, Russell's early intellectual development can serve to illustrate the continued cogency of the English view of mathematics well into the final decade of the nineteenth century. In some sense, the heat of his later scorn for that tradition can be taken as a measure of its importance for him at the time. Using Russell as a guide, then, this chapter will attempt to push past the wall of retrospective disgust which has been erected around late-century Cambridge mathematics in order to seen the final shape of the nineteenth-century English view of geometry. It will focus on the path which led, in 1896, to *An Essay on the Foundations of Geometry,* Russell's first published book. The intellectual development which culminated in this work illustrates the interaction of cultural, intellectual and foreign forces which were shaping the English geometrical tradition in the final decade of the century.

Russell's earliest statements about the epistemological status of

---

[3] Bertrand Russell, *My Philosophical Development,* (London: George Allen & Unwin, 1959), pp. 37–38.

[4] This statement cannot be made, however, without recognizing the existence of an alternative point of view, less ubiquitously expressed. See Karl Pearson, "Old Tripos Days at Cambridge, As Seen from Another Viewpoint," *The Mathematical Gazette,* 20 (1936):27–36.

non-Euclidean geometry are to be found in set assignments written after he had taken the Mathematical Tripos, in his final year as a student. The major ideas expressed in these short pieces represent the thinking of a brilliant, though very young, scholar exploring the parameters of a given philosophical issue. His views mirror the philosophical state of the problem in the final decade of the century. What is perhaps most striking about the picture they reflect is its stagnancy. Within philosophy, the English discussion of geometry had clearly stalled with the metric arguments of Helmholtz. Despite Cayley's 1883 plea that philosophers change their course, projective geometry played no role in Russell's philosophical ruminations.

Russell next wrote about geometry in the first half of 1895, when he penned a substantial manuscript entitled "Observations on Space." Although the "Observations" basically develops the same philosophical position Russell had sketched for his classes, it shows the influence of another tradition as well. Russell began composing it in Germany, and it reflects his exposure to continental approaches to geometry. In particular, Russell includes a brief consideration of projective geometry which he had discovered while on the Continent.

The culmination of this development was Russell's *Essay* itself, which he wrote at Cambridge in a fellowship year following his German trip. In this published work Russell's treatment differs from that in the "Observations" in significant ways. The new directions he here pursued point to the continued vitality of the English mathematical tradition which had thrived at Cambridge throughout the second half of the century.

RUSSELL AND NON-EUCLIDEAN GEOMETRY

Russell's earliest exposure to non-Euclidean geometry seems to have been before he began his studies at Cambridge. It is impossible to know exactly what he first read on the subject, but it was clearly material written in the spirit of the Ward-Stephen debate over necessary truth. Remembering the period sixty years later, he recalled:

> I discovered that, in addition to Euclidean geometry there were various non-Euclidean varities, and that no one knew which was right. If mathematics was doubtful, how much more doubtful ethics must

be! If nothing was known, it could not be known how a virtuous life should be lived. Such thoughts troubled my adolescence, and drove me more and more towards philosophy.[5]

Whatever his long-term educational goals might have been, when he entered Cambridge Russell was faced with the Mathematical Tripos. The immediate effect of preparing for this examination was to turn him resolutely away from mathematical study. He later remarked: "When I had finished my Tripos, I sold all my mathematical books and made a vow that I would never look at a mathematical book again."[6] Judging from the books from his library that remain in the Russell Archives at McMaster University, this statement appears to be literally true.

In his fourth year, having left mathematics behind him, Russell plunged into studies for the Moral Sciences Tripos. This led him directly into the philosophical world with which he had previously grappled only extracurricularly. In the 1890s the parameters of this philosophical world were very different from those defined by such British thinkers as Herschel, Whewell or even Mill. The philosophical background to Russell's development, therefore, requires a brief introduction.

One of the major unifying characteristics of the mid-century English philosophical position described in Chapter 1 was that scientific knowledge was treated as the norm of truth. The exemplar of true arguments or demonstrations was scientific. Truths, scientifically established, were unquestionable and known absolutely. In the 1860s and thereafter, however, developments within the sciences seriously undercut the comfortable world view which had been constructed on this philosophical basis. New scientific theories introduced a host of problems within the neat picture painted by men like Herschel, Whewell or even Mill. The most obviously disruptive development was the Darwinian theory of evolution by natural selection.

This theory seemed disturbing from virtually every perspective. It was methodologically suspect; the argument in Darwin's book did not conform to any of the theories of scientific methodology pro-

---

[5] Bertrand Russell, "A Turning-Point in My Life," *The Saturday Book*, 8 (1948): 143.

[6] Russell, *Development*, (1959), p. 38.

pounded in mid-century England. This meant in theory that the truth it offered was unsound, and yet it was so persuasive as to seem unanswerable. Furthermore, evolutionary theory was socially depressing; the most obvious model of human society which one could extrapolate from Darwin's theory was brutal and mean. Relatedly, it was theologically devastating; the picture Darwin drew of a random and bloody natural world was highly threatening to a theology which looked for the revelation of God's being in the carefully designed forms of His natural creation. On the platform of scientific philosophies like Herschel's and Whewell's, where science provided a norm of truth and a model of human aspiration, the new breed of scientific naturalists, like Huxley and Tyndall, relentlessly wielded Darwinian theory to undercut many cherished beliefs.

Evolutionary theory was the most obviously disturbing scientific development of the period, but there were many others. Non-Euclidean geometry is a good example. Yet others, like Maxwell's electromagnetic field theory, were also challenging. Each of these developments posed a variety of serious problems for the scientific philosophies of the mid-century. To quote a contemporary observer of philosophical development, none of the mid-century systems seemed adequate to deal with "the recent crowding of new scientific conceptions." None seemed "to present its leading principle bent as one would like to see it into the curves and junctures of the most anxious thought of our time."[7]

The massive problems new scientific developments posed for the scientifically grounded worldview of men like Whewell and Herschel led many intellectuals of the next generation to move in a radically different direction. Thinkers like Edward and John Caird in Scotland, James Ward and John McTaggart at Cambridge, and Thomas H. Green, Bernard Bosanquet and Francis H. Bradley at Oxford—though often very different—were united in turning away from philosophies which upheld science as the norm of truth. Instead, they embraced an idealistic stance, which emphasized that "the world with which science concerns itself is not fully real."[8]

---

[7] Quoted from David Mason (1875) in James Bradley, "Hegel in Britain: A Brief History of British Commentary and Attitudes," *The Heythrop Journal*, 20 (1979): 16.

[8] John Passmore, *A Hundred Years of Philosophy*, rev. ed. (New York: Basic Books, 1966), p. 64.

The philosophical alternatives which these men proposed were highly disparate and aimed at very different ends. Thus, for example, Green and his pupil Bosanquet were concerned with developing an approach to social thought which could act as a counterweight to untrammeled social Darwinism. Bosanquet, however, envinced very little interest in the religious questions which motivated much of Green's thought. Bradley, on the other hand, was also interested in Darwinian theory, but he found in it an ontological rather than theological or social challenge. Despite their differences, these men were both moving beyond the philosophical parameters in which the discussion of geometrical epistemology was joined and into idealistic approaches which emphasized the importance of realities which transcended scientific understanding. The ideas of Bradley, which strongly influenced the young Russell, can be used to illustrate the anti-scientific, in this case anti-mathematical, cast of idealistic thinking.

Bradley's *Principles of Logic,* published in 1883, marked the full flood of late nineteenth-century British idealism. One of Bradley's major interests in this work was the analysis of human thought. He categorically denied the relevance of any kind of formal system to this effort. Real reasoning, he argued, encompassed processes which simply could not be described by any such system. His example of a kind of reasoning which could not be formally described was relational reasoning of the kind "*A* before *B* and *B* with *C*, therefore *A* before *C*." In making this statement, Bradley was belligerently anti-mathematical. "If I knew perhaps what mathematics were," the Oxford-educated philosopher sneered, "I should see how there is nothing special or limited about them, and how they are the soul of logic in general and (for all I know) of metaphysics too." But, he continued, "Logic is not logic at all if its theory is based on a previous mutilation of the facts of the subject. It may do something which perhaps is very much better, but it does not give *any* account (adequate or inadequate) of reasoning in general."[9]

In its entirety, Bradley's argument is sophisticated and technical to a degree far beyond the level of this discussion. However, the basic thrust is familiar. It rests on essentially the same ground as

[9] F. H. Bradley, *The Principles of Logic,* 2 vols., (London: Oxford University Press, 1922), 1, p. 387.

Huxley's accusation that mathematics was empty and of no educational value. Similarly it reflects the themes of Wilson's attack on the artificiality of Euclidean geometry. In all of these cases, the attack on some form of mathematical argument involved isolating it from other forms of thought and turning it into an essentially meaningless play on words or symbols: a play which ultimately involved the manipulation of merely arbitrary conventions.

During his student years, Russell was converted to idealism and particularly impressed by the ideas of Bradley and McTaggart. However, unlike Bradley, he had prepared for the Tripos and was therefore not mathematically illiterate. Within the basic idealistic framework he embraced, Russell proposed a somewhat more complicated approach to the nature of mathematical knowledge. In his earliest works on the foundations of geometry, Russell developed a neo-Kantian view of the subject which was very close to the conceptual view of geometry embedded in Cambridge educational theory.

The first traces of Russell's view of geometry are to be found in a paper set for him in Ward's course on metaphysics. The assignment was

> Discuss a) meaning b) possibility of mentally representing other space-relations than those of Euclid.—Explain (look at Helmholtz) and discuss Helmholtz's distinction between geometry based on transcendental intuition and geometry based on experience. [Specific references were then made to the Helmholtz-Land papers.][10]

Russell opened his consideration of the nature of geometrical knowledge as follows:

> In Euclidean geometry certain axioms are assumed (such as especially the axiom of parallels) which depend upon the nature of space and which are held to derive their validity from the impossibility of picturing a case in which they fail. This impossibility is denied by Meta-Geometry.[11]

---

[10] Bertrand Russell, *The Collected Papers of Bertrand Russell*, ed. Kenneth Blackwell et al., vol. 1, (London: Allen and Unwin, 1983), p. 124.

[11] Ibid., p. 126. Here and elsewhere, Russell uses the term "meta-geometry" to mean non-Euclidean geometry.

He then considered whether Helmholtz had succeeded in enabling one to picture non-Euclidean geometries. Although he admitted that "if Helmholtz should say that he can picture them, I see no way of disproving his assertion," Russell did not accept the German's arguments as persuasive. He argued that the analytical demonstrations might establish the consistency of such spaces but not their imaginability. At the same time he brushed aside the more concrete examples, saying: "The analogy of flat-fish living on the surface of a sphere seems irrelevant, if only because we are not flat-fish, and I do not see why the line of sight of such a flat-fish should not be just as much a Euclidean straight line as ours." He concluded with the comment:

> I do not see how we can avoid the conclusion, that geometry, to be geometry, . . . must depend ultimately on space-intuitions . . . , further that such space intuitions are *for us* necessarily Euclidean; and that therefore the speculations of meta-geometry have no epistemological importance.[12]

This conclusion, in which Russell emphasized the inadequacy of the formal structure of non-Euclidean or meta-geometry to capture the essence of intuitively known space, put him comfortably into the idealist camp. It can be seen as reflecting Bradley's position that scientific or mathematical knowledge was not relevant to knowledge of the real or Absolute. Like Bradley, Russell affirmed the primacy of a kind of knowledge which transcended the attempts of formal scientific argument to capture it.

In his emphasis on transcendent truth, though, Russell's position is equally reflective of the long British tradition which had informed his geometrical education. His insistence on the primacy of spatial understanding over mathematical forms echoes Jevons as much as Bradley, Cayley as much as Ward. It seems unlikely that at this early date Russell knew of the geometrical ideas of either Cayley or Jevons, or that he was familiar with most of the rest of the British corpus on non-Euclidean geometry. Nonetheless, because he viewed geometrical forms ultimately as attempts to describe a con-

---

[12] Ibid., p. 127.

ceptual entity, his nascent philosophical ideas were neatly in line with geometrical attitudes which had informed the mathematical education he had received at Cambridge.

The next stage of Russell's geometrical development was set in Germany, where he went after finishing Cambridge. Russell spent the first three months of 1895 in Berlin studying economics. During this period, he recalled his "intellectual ambitions were taking shape," and he imagined himself writing "one series of books on the philosophy of the sciences from pure mathematics to physiology, and another series of books on social questions."[13] The first of these works was to be on geometry. While still in Berlin, Russell began a manuscript, which he finished in June and entitled "Observations on Space and Geometry."

Russell's "Observations" is much longer and more detailed than the set pieces he had written for Ward's course. In an "Introduction" appended to the manuscript, Russell presented a neo-Kantian thesis. Our conception of space, he argued, is partially subjective and partially objective. Furthermore, in its most primitive form, space is an amorphous intuition as opposed to a structured conception. The process of generating a full-blown conception itself requires considerable "criticism and reflection." These psychological processes, he continued, considerably muddy any attempt to clearly understand the ultimate components of the spatial concept. "Hence, meta-geometry and the a priori of space are not obviously incompatible—and neither can, without further investigation, be regarded as an argument against the other."

Russell described the relationship between meta-geometry and the spatial concept in two numbered propositions:

> 1) That even if sensational space were subjective, it would not afford sufficient ground for an objection to Metageometry, and vice versa . . . 2) That Geometrical Axioms owe their origin in part only to spatial intuition, the remainder being due to purely logical motives

---

[13] Bertrand Russell, *The Autobiography of Bertrand Russell, 1872–1914,* (London: Allen and Unwin, 1967), pp. 184–85. For a full discussion of the shape of Russell's early intellectual program, see Nicholas Griffin, "The Tiergarten Programme," *Russell* (forthcoming).

. . .—that the latter element, like Arithmetic, has apodeictic certainty while the former is only approximate and empirical.[14]

The first of these statements seems to point towards a strict separation of geometrical conception and mathematical development. The second assigns to geometry a dual nature, suggesting an interesting and complex intermediate position between a strictly formal or definitional view and a descriptive or conceptual one. The rest of the "Observations" is not clearly structured around demonstrating these assertions, but it sheds some light on what Russell might have meant by them.

Russell pursued his investigation into the nature of the spatial conception and its relation to geometry, first by looking at philosophical works, then at mathematical ones. His "Observations" began with a detailed criticism of the work of two German metaphysicians, Heinrich Lotz and Franz Erhardt. Both of these men had defended neo-Kantian conceptual views of space against the empirical claims of Helmholtz and his followers. Russell basically agreed with their position that the space described by Euclidean geometry was an *a priori* concept whose essential epistemological status was unaffected by non-Euclidean developments. He could not resist displaying his sharp critical skills and mathematical prowess, however, and devoted the first third of his manuscript to a detailed critique of their mathematical argument. Russell then proceeded to a historical treatment of the nineteenth-century development of non-Euclidean geometry. Here he directed the bulk of his attention to the work of Riemann and Helmholtz. Whereas he had taken Lotz and Erhardt to task for their mathematical weaknesses, Russell attacked Riemann and Helmholtz for their philosophical ones.

Russell focused first on Riemann. The mathematician, he acknowledged, had done a masterful job of reducing the axioms of space to their barest analytical roots. However, the philosophical

---

[14] Bertrand Russell, "Observations on Space and Geometry. Berlin, March, 1895," in The Bertrand Russell Archives, McMaster University, Hamilton, Ontario, p. 3. (All quotations are taken from a typescript of this work, which is in the archive. Page numbers refer to the typescript pages. Except for the "Introduction" for which they are unavailable, the manuscript pages are given in parentheses.)

problem, as Russell saw it, was how this reduction was related to the spatial *concept*. In this context, Russell pointed out, Riemann's work is of doubtful relevance because "we have, throughout the argument, no point of contact with actual perception."[15] He summed up his remarks by saying:

> Mathematically, Riemann's form is probably as good as any that can be imagined; but philosophically it seems to me very ill fitted to settle what space-conception we require to fit our space-perceptions; and this is the question on which turns the truth to fact of any Geometry, as opposed to mere logical self-consistency.[16]

Russell had a different problem with Helmholtz, because the physiologist had directly linked the development of Euclidean concepts to common experience. In response to this direct challenge, Russell, like Land, focused attention on Helmholtz's arguments for the conceptual equality of Euclidean and non-Euclidean spaces. Helmholtz's elaborations of the experiences one would have in non-Euclidean space, Russell argued, served merely to describe these hypothetical spaces; it did not render them imaginable and hence real.

> We really derive no more knowledge of it [non-Euclidean space] from this analogy, than a man born blind may have of light; he may be perfectly well-*informed* on the subject, and may even be able, theoretically, to work out mathematical problems in Optics; but he cannot, in any ordinary sense of the word, *imagine* light—he has the *conception*, but lacks the living *intuition*.[17]

Whereas Russell criticized Riemann's work as too abstract for epistemological relevance, Helmholtz's was too concrete. Thus, albeit for different reasons, he concluded that both of their works were ultimately irrelevant to the philosophical status of the spatial concept.

---

[15] Ibid., p. 34 (65).
[16] Ibid., p. 35 (67–8).
[17] Ibid., p. 43 (85).

Russell's early defenses of the *a priori* nature of spatial knowledge were thus argued largely within the confines of the issues arising out of the metrical geometry of Riemann and Helmholtz. Although the second of the points in his introduction might suggest that he would find the formality of their analytical treatments attractive, Russell seems to have emphasized the first point. He dismissed the metageometrical ideas of Riemann and Helmholtz as essentially irrelevant to understanding the conception of space which stood transcendent, totally beyond the reach of either analytical or experiential reasoning.

RUSSELL AND PROJECTIVE GEOMETRY

From a mathematical perspective, what is striking about Russell's early treatments of non-Euclidean geometry is how narrowly he confined his attention to the issues raised by Riemann and Helmholtz. This consistency of focus was maintained even though in the historical section of his Berlin "Observations" Russell emphasized that in mathematics, after the early 1870s, the metrical ideas which underlay Riemann's and Helmholtz's analysis of space had been largely abandoned in favor of a projective approach.

It seems that Russell had only recently learned of projective geometry when he wrote his "Observations." He had renounced his mathematical education at Cambridge before taking the second part of the Tripos, which would have included the subject.[18] Thus, he seems to have first studied projective geometry in Germany. The text he studied was Klein's *Vorlesungen über Nicht-Euclidische Geometrie,* which he praised as "one of the very best textbooks I have ever come across."[19]

The view of projective geometry Russell gleaned from Klein's *Vorlesungen* was very different from the quasi-inductive English view, described above in Chapter 3. Although both continental and

---

[18] For a detailed treatment of the structure of the Tripos when Russell took it, see W. H. Besant, "The Mathematical Tripos," *The Student's Guide to the University of Cambridge,* 5th ed. (Cambridge: Deighton, Bell and Co., 1893).

[19] Russell, "Observations," (1895), p. 57 (121).

English mathematicians eagerly followed Klein's lead and pursued projective geometry in the final quarter of the century, the two groups developed highly divergent interpretations of its meaning. In England, the interpretation was highly descriptive and often grounded in some form of the principle of continuity. On the continent, on the other hand, the foundations of projective geometry were increasingly seen to lie in the abstract theory of "throws" developed by Karl von Staudt in his *Geometrie der Lage* of 1847.[20]

Although Russell was clearly fascinated by the subject, the formally based development he found on the continent had little relevance to his philosophical interest in the *a priori* bases of the spatial concept. In the "Observations," Russell characterized the change in mathematical emphasis engendered by projective geometry as follows:

> The fundamental distinction, between this [projective period] and the preceding [metric] period, is one of method alone—whereas Riemann and Helmholtz dealt with metrical ideas, and took as their foundations the measure of curvature and the formula for the linear element (both purely metrical), the new method takes at is foundations the formulae for transformation of coordinates, which are required to express any given motion, and begins by reducing all so called metrical notions (distance, angle, etc.) to projective forms. The treatment derives, from this reduction, a methodological unity and simplicty [sic] before impossible.[21]

Clearly Russell saw projective geometry as mathematically attractive because it provided an elegant and powerful analytical method to solve geometrical problems.

Philosophically, however, it was virtually meaningless, a point Russell emphasized in the sentence preceding the above quotation. "Of this [projective] period," he wrote, "I can only treat with great brevity, as there are few fresh philosophical ideas involved; the advance is mainly technical, and impossible to render in non-math-

---

[20] Karl Georg Christian von Staudt, *Geometrie der Lage*, (Nürnberg: F. Korn, 1847). This is the point of view taken by Felix Klein in *Vorlesungen über Nicht-Euklidische Geometrie* (Göttingen, 1982), Chapters I & II. Russell explicitly refers to these chapters in his "Observations," p. 52(106).
[21] Ibid., p. 50 (103–104).

ematical terms."[22] By interpreting projective geometry as a method, Russell distanced it from the fundamental spatial concept which formed the object of his philosophical interest.

Russell was not completely comfortable dismissing projective geometry in this way, however. "Nevertheless," he continued, "I will give a brief sketch of the two most important advances made by it, namely Cayley's projective Geometry and Lie's treatment of motions; . . . at the same time I will endeavour to point out certain philosophical difficulties in the former."[23] He then launched into a fifteen-page discussion of the subject.

Russell's treatment of Lie's ideas was fairly technical and directed towards a weak point in Helmholtz's mathematics which Lie had discovered. When he turned to the English tradition led by Arthur Cayley, however, Russell was struggling with the broader issues of spatial conceivability. In a sense, his Berlin "Observations" was the first serious attempt to respond to the philosophical challenge Cayley had made in his BAAS address of 1883.

Russell's response to Cayley's specific concern with the conceptual status of imaginary points or points at infinity was dismissive. Their central role in Cayley's projective theory of distance served merely to emphasize for Russell the irrelevance of that theory to the essential spatial concept. In his words:

> for philosophical purposes the reduction [of metrical to projective space through the Cayley-Klein theory of distance] is irrelevant, since it depends, usually, upon imaginary points and figures . . . , which have no reality apart from symbols.[24]

Russell devoted more energy to the conceptual problems raised in Cayley's "Note" appended to the "Sixth Memoir": the nagging question about whether the projective theory of distance was circular. "The difficulty is," Russell explained, "that the values of our coordinates already involve the usual measure of distance, so that to give a new definition while retaining the usual coordinates is to

---

[22] Ibid.
[23] Ibid., p. 51 (104).
[24] Ibid., pp. 50–51 (104).

incur a contradiction."[25] Russell's response was primarily aimed at the view of this problem developed by the Irish astronomer Robert S. Ball.

Professionally, Ball was an astronomer, but in the nonspecialized English tradition, this label certainly did not circumscribe his interests. One of them, as was mentioned above, entailed exploring the dynamics of moving bodies in non-Euclidean spaces. In addition, he was drawn towards the epistemological challenges involved in non-Euclidean developments. He brought a sophisticated understanding of projective geometry to this problem, which pulled his musings out of the metric tradition established by Helmholtz. His work is a unique and important statement about the nature of geometry, which grew from within the English projective tradition.

Ball's earliest presentation of non-Euclidean geometry, written in 1879 for a wide audience, provided evidence for the mathematical community's shift towards interpreting the new work projectively. In this paper Ball had relied on Cayley's four-point definition of distance in order to present metric geometry within a projective framework. However, as was pointed out in Chapter 3, that discussion hinted that he was troubled by the conceptual demands attendant on accepting that distance really was a relationship which involved not only two perceived points, but two imaginary ones as well. "It cannot be denied that there *appears* to be something arbitrary in this definition," he wrote.[26] But, at that time, despite his discomfort, he pushed ahead and acted on the assumption that Cayley's definition of distance was conceptually valid.

In 1887, however, Ball radically altered his approach. In a paper entitled "On the Theory of the Content," he returned to consider the problem he had glossed over before. His approach was completely different, however; he abandoned the attempt to conceptually ground the mathematical theory which was demanded by a descriptive notion of rigor. Instead, he unabashedly tried to split mathematical forms from their conceptual contents, a move which

---

[25] Ibid., p. 52 (107–108).
[26] Robert S. Ball, "The Non-Euclidean Geometry," *Hermathena*, 3 (1877–79): 502. The full quotation is on pp. 154–155.

implied a different notion of mathematical truth—a notion based on consistency rather than conceptual clarity.

Ball directly addressed his "Content" paper to the conceptual issues raised by Cayley's treatment of distance.

> In the study of the so-called 'Non-Euclidean Geometry' I have often felt a difficulty which has, I know, been shared by others. In that Theory it seems as if we try to replace our ordinary notion of distance between two points by the logarithm of a certain anharmonic ratio. But this ratio itself involves the notion of distance measured in the ordinary way. How, then, can we supersede our old notion of distance by the non-Euclidean notion, inasmuch as the very definition of the latter involves the former?[27]

Ball tried to circumvent this conceptual problem, which was stated in terms of "notions," by developing a nonconceptual, or formal, mathematical structure. He proceeded to elaborate a theory of the "content" which was self-consciously free of spatial language. He generated a series of definitions for the elements of this nonconceptual entity: its "objects," their "ranges," their "extents" and the "intervene." He introduced a number of axioms which governed their relations. Only after he had elaborated an entire system and derived a number of theorems did Ball attach meaning to his terms by matching the elements of his content with those of space—points with objects, lines with ranges, planes with extents and distance with the intervene.

Ball first presented his theory of the content abstractly, in an attempt to remove the mathematical discussion of distance entirely from the conceptual sphere. He emphasized that his intervene was defined wholly by the axioms and other elements of the system and thus ought to carry no intuitive baggage.

> The theory of the content is no doubt equivalent to . . . the theory of so-called non-Euclidean geometry or geometry of elliptic space. I have indeed, myself, frequently used these latter expressions, but am

---

[27] Robert S. Ball, "On the Theory of the Content," *Transactions of the Royal Irish Academy,* 29 (1887–89):123.

glad to take this opportunity of renouncing them. I have found the
associations they suggest very misleading; and, in fact, it was only by
employing some such conceptions and terminology as that used in
the present Paper that I succeeded in understanding the subject. *It
is especially needful to avoid confusing the sign with the thing signified.*
The space-points are only the signs of the objects—they are not the
objects themselves—and, if we are not careful to preserve this distinction, obscurity will arise about the 'Intervene' [that is, distance].[28]

This statement of general principle represents a real departure from
the descriptive view of mathematical rigor, wherein the mathematical object and its meaning were essentially inseparable. Ball's insistence on separating the mathematical sign from the conceptual
thing signified was just the separation which had been resisted by
such mid-century philosophers as Whewell, Herschel or even Mill.
It thrust mathematics out of her position as the epistemological
Queen of the Sciences by breaking her ties with others forms of
knowledge. On the other hand, it allowed Ball to skirt the conceptual quagmire of Cayley's theory of distance with the clarity of mathematical forms.

Perhaps because Ball's message was so radical, it seems to have
been hard for his contemporaries to follow it. Cayley referred to
Ball's paper in the final sentence of his "Note" to the "Sixth Memoir" but only to say, "In it a similar problem is treated."[29] This brief
comment seems to be what first led Russell to consider Ball's paper
in his "Observations." There Russell acknowledged that the theory
of the content allowed Ball to clear up the nagging conceptual problem associated with the definition of distance in projective space.
"Sir R. Ball's account is probably the only tenable one. But," he
continued, "with the confusion, the supposed Euclidean interpretation also disappears; Sir R. Ball's Content, if it is to be in any sense
a *space,* must be a space radically different from Euclid's."[30]

In this case, as it was with Riemann's work, Russell's position
seems to have been that while formal mathematical developments

---

[28] Ibid., p. 151.

[29] Arthur Cayley, "Notes and References," *The Collected Mathematical Papers of Arthur Cayley,*
11 vols., (Cambridge: University Press, 1893), 2, p. 605.

[30] Russell, "Observations," p. 53(110–11).

and models could be invaluable in clearing up logical difficulties in mathematical arguments, they had no bearing on philosophical issues concerned with the spatial concept.

Clearly, in 1895 Russell did not see that projective geometry could have any relevance for the epistemological questions posed by the development of non-Euclidean geometry. Instead, his philosophical treatment of the *a priori* spatial concept remained rooted in metrical conceptions. The 1897 *Essay*, on the other hand, reveals a dramatic shift in this defense. Here Russell defended the same philosophical position from a radically different standpoint, by placing projective geometry in a central position. Projective space became an integral part of the *a priori* spatial concept. The *Essay* brings the essential thesis of the "Observations" more closely into line with developments in the conceptual British mathematical tradition that he had essentially ignored in his earlier work.

AN ESSAY ON THE FOUNDATIONS OF GEOMETRY

There are few suriving manuscripts relevant to the development of Russell's spatial ideas in the two years separating the "Observations" from the published *Essay*. In August of 1895, he submitted to Cambridge a fellowship dissertation on the foundations of geometry. This manuscript has since been lost. We know only what Russell later reported: that Alfred North Whitehead, the mathematical reader, "criticized it rather severely, though quite justly, and I came to the conclusion that it was worthless."[31]

The specifics of Whitehead's objections are lost in the mists of time. Whatever they were, Russell was awarded the fellowship and, despite his initial discouragement, did not abandon the subject. In the "Preface" to his published monograph, *An Essay on the Foundations of Geometry,* Russell states that it is a revised version of the now missing manuscript. In the absence of the dissertation itself, or a fuller description of the objections raised to it, Russell's development can best be traced by comparing the *Essay* with the "Observations."

---

[31] Russell, *Autobiography,* (1967), p. 186.

From his perspective, Russell's *Essay* was the first finished work in the neo-Hegelian program he had set for himself while in Germany. The opening words of the work itself, however, seem to place his concerns firmly within the English tradition in which he had been raised.

> Geometry, throughout the 17th and 18th centuries, remained, in the war against empiricism, an impregnable fortress of the idealists. Those who held—as was generally held on the Continent—that certain knowledge, independent of experience, was possible about the real world, had only to point to Geometry: none but a madman, they said, would throw doubt on its validity, and none but a fool would deny its objective reference. The English Empiricists, in this matter, had, therefore, a somewhat difficult task; either they had to ignore the problem, or if, like Hume and Mill, they ventured on the assault, they were driven into the apparently paradoxical assertion that Geometry, at bottom, had no certainty of a different *kind* from that of Mechanics—only the perpetual presence of spatial impressions, they said, made our experience of the truth of the axioms so wide as to *seem* absolute certainty.[32]

The parallels between this passage and the one with which Russell's contemporary, Dixon, had opened his discussion of geometry five years before is striking.[33] Though Russell's and Dixon's educational trajectories were close during this period, there is no sign that Russell noticed Dixon or his work at all. The similarity of their approach attests to the power of the English tradition in geometry in which each of them had been steeped rather than any direct influence.[34]

The first chapter of Russell's *Essay* is historical and borrows heavily from the "Observations." There are, however, some significant departures from the earlier work, particularly with respect to pro-

---

[32] Russell, *An Essay on the Foundations of Geometry*, (Cambridge: University Press, 1897). p. 1.
[33] See the passage quoted on pp. 186–187.
[34] Russell ignored Dixon, but Dixon did not ignore him. Dixon's third book was a detailed review of Russell's *Essay*, first delivered to the Aristotelian Society on December 13, 1897, and then published as a monograph entitled *A Paper on the Foundations of Projective Geometry*, (Cambridge: Deighton, Bell and Co., 1898). Apparently Russell did not respond to this work, any more than he did to Dixon's earlier efforts.

jective geometry. In the 1897 work, rather than asserting the philosophical irrelevance of projective ideas, Russell stated,

> The third [projective] period differs radically, alike in its methods and aims, and in the underlying philosophical ideas, from the period which it replaced. Whereas everything, in the second [metric] period, turned on measurement . . . these vanish completely in the third period, which, swinging to the opposite extreme, regards quantity as a perfectly irrelevant category in Geometry. . . . The ideas of this period, unfortunately, have found no exponent so philosophical as Riemann or Helmholtz, but have been set forth only by technical mathematicians.[35]

Russell's *Essay* can be seen as an attempt to provide the philosophical treatment of projective geometry which had been lacking so long. In this, which Russell effected at a high technical level, he can be seen as bringing fruit out of a fertile combination of his mathematical and philosophical knowledge and acumen. The mathematical details of his construction are grounded in an impressive understanding of both projective and metric geometries as his contemporaries were pursuing them.

Philosophically, Russell's work was highly sophisticated. With it he can be seen as having made a significant contribution to the late nineteenth-century philosophical discussion of the nature of geometry. He can equally well be seen as consolidating and reinterpreting the informal philosophical orientation from which English mathematicians had been pursuing projective ideas since the 1860s. The basic structure of his argument, in which projective space forms the *a priori* substructure for Euclidean space, reflects, albeit in a somewhat distorted form, the English perspective which had first emerged in the 1870s.

In the *Essay*, Russell's defense of the *a priority* of the spatial concept began with the recognition that projective geometry involved the development of fundamental spatial ideas. In this sense he moved away from his continentally formed earlier view that the study merely represented a new method with which to approach classical Euclidean space towards the more English view which iden-

---

[35] Russell, *Essay*, (1897), pp. 27–28.

tified projective geometry with its own space. As he did this, Russell was less naïvely concrete than many of the British mathematicians who had preceded him. He did not insist that each projective entity have a real spatial interpretation. Instead, he asserted that the qualitative conceptions lying behind the mathematical forms of projective space were *a priori*. This gave him some freedom to approach those forms in new ways.

In fact, Russell's late-century defense of the peculiarly English view that projective geometry described a conceptually real projective space entailed radically attacking a number of the basic points on which that view had previously rested. In particular Russell's construction entailed dismissing imaginary points as mere analytic artifacts. More fundamentally, it required not identifying but strictly *separating* notions of distance as they appeared in projective and metric geometries, in order that they not be considered arbitrarily interchangeable. Russell's treatment of these issues clarifies the basic parameters of his philosophy of space.

Russell's consideration of projective *space* led him directly into the question of the reality of the portions of that space which had hitherto been so problematic, notably the imaginary points. Whereas in the "Observations" he had considered the reality of these points only in order to substantiate the philosophical irrelevance of projective geometry, in the *Essay*—where he was taking projective space seriously—he had to consider their ontological status directly. He treated them equally dismissively, however. "As well might a postman assume that, because every house in a street is uniquely determined by its number, therefore there must be a house for every imaginable number,"[36] he scoffed. Imaginary points and points at infinity instances of numbers without houses; they were mere analytical artifacts. With this bald assertion, Russell dismissed them entirely.

Russell's approach to the problem of distance involved an even more radical break with English tradition. The issue which refocused Russell's attention seems to have been conventionalism. This view was being championed by Henri Poincaré, a French polymath, equally conversant with mathematical and philosophical

---

[36] Ibid., p. 44.

ideas. Early in the final decade of the century, Poincaré began publishing a series of articles which eventually formed the core of the geometrical discussion in his book *Science and Hypothesis*.[37] An early one of these papers, translated and published in *Nature* in 1892, seems to have alerted Russell to serious problems with the dismissive treatment of projective geometry he had given in the "Observations."

In the *Essay*, Russell presented the conventionalist position as follows:

> 'What ought one to think,' he [Poincaré] says, 'of this question: Is the Euclidean Geometry true? The question is nonsense.' Geometrical axioms, according to him, are mere conventions: they are 'definitions in disguise.'[38]

Poincaré had argued this point of view in the context of metrical geometry. He had interpreted the various consistency models which allowed one to treat non-Euclidean geometries in Euclidean terms as "translations." When viewed in this manner, the differences among Euclidean and non-Euclidean geometries had no more intrinsic meaning than did differences among languages; as languages expressed a common meaning, the various geometries merely represented different conventional systems for dealing pragmatically with a common spatial reality. The implication of this position was that there were no essential differences among geometrical systems and, consequently, no geometrical system could capture a better or more true conception of space than any other.

Russell's rejection of conventionalism had been foreshadowed in his "Observations." There, he briefly ascribed a conventionalist view to Klein, who in Russell's words "tends to regard the whole controversy [about the truth of geometrical systems] as a mere question as to the definition of distance, not as to the nature of space— for by mere alterations in this definition we can, from a Euclidean

---

[37] *Science and Hypothesis*, along with Poincaré's other major philosophical works, *The Value of Science* and *Science and Nature*, have been collected in one volume: *The Foundations of Science*, auth. trans. George Bruce Halsted, (Washington, D.C.: University Press of America, 1982).
[38] Russell, *Essay*, (1897) p. 30. Russell's quotations are from Poincaré, "Non-Euclidean Geometry," trans. W. J. L., *Nature*, 45 (1892):407.

plane, derive all these different systems."[39] Russell countered this claim by denying that the "space" in which these definitions were being introduced was truly a space at all—being analytically defined, it had no legitimate spatial referent.[40]

In the *Essay*, where he was treating projective space as the fundamental underpinning of the spatial concept, however, Russell had to develop a different response to the conventionalist argument. The projective treatment of non-Euclidean geometry Cayley had developed seemed highly vulnerable to conventionalist interpretation. If the various metrical geometries could be obtained merely by varying arbitrary definitions of distance in an *a priori* projective space, the choice among them could easily be seen as meaningless and conventional. This possibility simply had not occurred to Cayley, who staunchly maintained the *a priori* truth of Euclidean geometry throughout this mathematical career.

Russell met this challenge by assigning *a priority* to distance. He argued that distance was a relationship between two points, not among four, and that the whole study of metrical geometry was intricately intertwined with the development of this basic idea. Projective geometry, on the other hand, studied the qualitative features of space without reference to the quantitative notion of distance at all. It was possible to work with coordinates in this system, defining them wholly qualitatively or projectively, as von Staudt had. Since projective coordinates were strictly qualitative, they bore no relation to real distance. Although functions could be defined on them which had the mathematical properties of quantitative distance, these could never be properly "distance" functions.

The feature supporting this radical separation lay in the fact that projective distance was a relationship among four points rather than between two. Cayley and those who followed him had interpreted this to indicate the inadequacy of the two-point relationship; Russell, on the other hand, firmly defended the two-point relationship and instead denied the legitimacy of identifying it with its four-point projective counterpart. He argued that by naming functions "distance" because they had algebraic properties *like* those of metrical

---

[39] Russell, "Observations," (1895), p. 52 (107).
[40] Ibid., p. 53 (110–11).

distance, projective geometers had created great confusions for themselves. Their apparently metrical relations were not true distance, but only "conventional symbols for purely qualitative spatial relations." To quote from Russell's explanation of this view:

> Distance in the ordinary sense is ... that quantitative relation, between two points on a line, by which their difference from other points can be defined. The projective definition, however, being unable to distinguish a collection of less than four points from any other on the same straight line, makes distance depend on two other points besides those whose relation it defines. No name remains, therefore, for distance in the ordinary sense, and many projective Geometers, having abolished the name, believe the thing to be abolished also, and are inclined to deny that *two* points have a unique relation at all. This confusion, in projective Geometry, shows the importance of a name, and should make us chary of allowing new meanings to obscure one of the fundamental properties of space.[41]

In this argument, the connection between qualitative, projective space and quantitative, metrical space suggested by the Cayley-Klein theory of distance was only apparent because the distance between two points was a fundamental metrical property only palely imitated by the analytic manipulation of cross ratios. In the "Observations," this line of reasoning had led Russell to reject the *spatial* relevance of projective geometry; in the *Essay,* where he took the space that geometry defined as *a priori,* he used it to argue that distance could not be properly defined in that space. Thus, Russell renegotiated the mathematical and philosophical matrix of the 1890s in order to defend the basic view which had been a mainstay of English work for so long: that geometry, including projective geometry, was a conceptual study.

It was a somewhat strained defense, however, because it assigned geometry a double character. With it, Russell separated much of what passed as geometry from its spatial referent while leaving other parts strictly connected. The whole notion of qualitative projective

---

[41] Russell, *Essay,* (1897) p. 36.

space was kept as real and judged to represent a significant conceptual advance over what had gone before. On the other hand, the imaginary points and distance of projective geometry became mere analytical artifacts or accidents of language. Some of geometry remained the necessary study of space, but much of it was merely technical analysis, developed separately from space or any other conceptual reality.

Because he maintained a neo-Kantian focus on conceptual space, Russell's *Essay* is less of a departure from the English geometrical tradition than Dixon's or Ball's works were. This conceptual feature saved him from the difficulties both of the others had faced with trying to explain what would constitute a non-conceptual form of reasoning. If geometry were really a merely formal study, how could one move from theorem to theorem? Without concepts to work with, how could one think at all? Dixon's reviewers criticized him particularly effectively on this point. Ball, whose paper elicited minimal response, was also caught on it. The first part of "On the Theory of the Content" was devoted to the formal theory; in the second part, however, Ball returned to the spatial language he had previously rejected in order to generate new knowledge about the content.

The conceptual aspects of Russell's account saved him from these difficulties but, on the other hand, kept him embroiled in the kind of conceptually muddy problems which had plagued his predecessors. Though they were carried out at a higher level of mathematical and philosophical sophistication, Russell's arguments in papers like his 1899 "Sur les Axiomes de la Géométrie" call to mind the kind of conceptual mire in which Dodgson's characters were trapped as they discussed adjustments in elementary textbooks. Russell closed this article, which was composed as an attempt to answer Poincaré's objections to his *Essay*, with the following expression of confusion:

> I have done my best to respond to the questions Mr. Poincaré indicated as the most important. I cannot hope that all of my responses will satisy him, and I fear that some will seem to him to be more or less irrelevant. I have done my best to sincerely consider his objec-

tions, and I would regret it more were I to fail in this regard than in any of my other intentions.[42]

It appears that in his private efforts to hammer out a single conceptual treatment, Russell encountered the same kinds of frustrations educators had met in their attempts to do the same as a group.

Thus, the development of the geometrical ideas which appeared in Russell's *Essay on the Foundations of Geometry* took place in several intellectual traditions. The earliest and, throughout, the strongest influence was the English epistemological tradition in which geometry was the descriptive study of an independently defined, conceptual subject matter. This tradition both informed the intellectual culture in which Russell located his first interest in the epistemological status of non-Euclidean geometry and shaped his formal mathematical education as a Cambridge undergraduate. Although the inadequacy of that education is a persistent theme in his retrospective evaluations of his intellectual development, his early work bore the mark of its philosophical underpinning in fundamental ways.

In the year after he finished Cambridge, Russell enthusiastically worked in a second tradition, German geometry. Despite his consistent retrospective praise for German as opposed to English mathematics, in his philosophical works written during this period the themes of his English training emerge and temper his evaluations of the continental tradition. Upon his return to England to revise his fellowship dissertation into a book, these themes become even more pronounced. In fact, much of the distinctive character of the *Essay* derived from Russell's understanding of his English mathematical heritage.

This point was clearly recognized by Russell's most enthusiastic French reviewer, Louis Couturat. In lauding Russell's work, Cou-

---

[42] Bertrand Russell, "Sur les Axiomes de la Géométrie," *Revue de Métaphysique et de Morale*, 7 (1899):707. In the original French, the passage reads: "J'ai essayé répondre de mon mieux aux questions indiquées par M. Poincaré comme les plus importantes. Je ne puis espérer que toutes mes réponses lui paraissent satisfaisantes, et je crains que quelques-unes ne lui semblent plus ou moins incongrues (irrelevant). J'ai fait mon possible pour tenir sincèrement compte de toutes ses objections, et je regretterais bien plus d'avoir failli à cet égard que d'avoir échoué dans toutes mes autres intentions." (The translation is mine.)

turat did not allow his praise to fall wholly on Russell as an individual. He placed the credit largely on the English education Russell
had received.

> It is permissible to regret but not to be astonished that such a spirit
> is not to be found in France: the fault lies not with people but with
> institutions, with the absurd system of bifurcation which continues to
> rule in the organization of our studies, and the deplorable split which
> results between philosophy and the sciences which are her necessary
> sustenance. Thus, the honor of bringing the discoveries and progress
> of modern geometry to bear on the theory of knowledge goes to an
> Englishman.

Unfortunately, although in correspondence Coururat specifically
asked Russell to comment on his English education, we do not have
Russell's reply.[44] It would be interesting to know his immediate
assessment of that education, faced with such an enthusiastic
observer.

In an important sense, then, the *Essay* can be seen as the culmination of the late nineteenth-century English geometrical tradition.
In it Russell renegotiated the basic themes of descriptive geometry
and conceptual truth which had defined the parameters of the
English approach since the middle of the century. He defended a
transcendentally true geometry against the whole panoply of objections which had been raised by new developments in the preceding
fifty years.

Russell's *Essay* was an impressive piece of work. With some of the
kind of retrospective humor about himself which contributes to his

---

[43] Louis Couturat, "Essai sur les fondements de la géométrie par Bertrand Russell," *Revue de Métaphysique et de Morale*, 6 (1898):354–55. In the original French, the passage reads: "Qu'un tel espirt ne se soit pas rencontré en France, il est premis de le regretter, mais non de s'en étonner: la faute en est, non aux hommes, mais aux institutions, à cet absurde système de la bifurcation qui continue à régner dans l'organisation de nos études, et à la déplorable scission qui en résulte entre la Philosophie et les connaissances scientifiques, qui en sont l'aliment nécessaire. C'est donc à un Anglais qu'était réservé l'honneur de résumer et de tirer au clair les découvertes et les progrès de la Géométrie moderne [projective geometry], et d'en prof- iter la Théorie de la connaissance." (The translation is mine.)

[44] Louis Couturat, Letter to Bertrand Russell, Oct. 3, 1897 in Bertrand Russell Archives, McMaster University, Hamilton, Ontario.

charm, Russell later recalled: "I had when I was younger–perhaps I still have–an almost unbelievable optimism as to the finality of my own theories. I finished my book on the foundations of geometry in 1896, and proceeded at once to what I intended as a similar treatment of the foundations of physics, being under the impression that the problems concerning geometry had been disposed of."[45] Despite his confidence, the interpretation he there developed offered an intrinsically unstable solution to the problems he was addressing. Convinced that geometry was ultimately grounded in spatial concepts, Russell nonetheless tried to free some of it from the strictures such a grounding imposed. He allowed that some elements of the study, like imaginary points, were merely analytic, while at the same time retaining the advantages of the old notion of spatial truth. He apparently did not see that the arbitrary separation between mathematical and conceptual development he was proposing might create strains which would pull the two entirely apart. Yet this was to be the particular fate of Russell himself, and the larger fate of the English intellectual tradition in which he was working. In the following decade Russell and Whitehead both abandoned the descriptive view of geometry defended in the *Essay* and together moved wholeheartedly into the formal development of logistics. Even as they moved in this direction, the institutional and intellectual tradition which had so long supported descriptive geometry in England faded away.

---

[45] Russell, *Development*, (1959), p. 41.

G. H. Hardy. *(Courtesy of the Master and Fellows of Trinity College Cambridge.)*

The Demise of English Descriptive Geometry

Russell's *Essay* can be read as the philosophical apex of the nineteenth-century English geometrical tradition, the international intellectual interest it generated as a tribute to the enduring strength of the conceptual approach to geometry. It was a short-lived triumph, however. Russell repudiated the book soon after he wrote it; "apart from details," he later wrote, "I do not think that there is anything valid in this early book."[1]

There is no clearer witness to the importance of this intellectual conversion than Russell himself.

> There is one major division in my philosophical work: in the years 1899–1900 I adopted the philosophy of logical atomism and the technique of Peano in mathematical logic. This was so great a revolution as to make my previous work, except such as was purely mathematical, irrelevant to everything that I did later. The change in these years was a revolution; subsequent changes have been of the nature of an evolution.[2]

By 1900 he had turned from the conceptual ideas on which the *Essay* rested and moved into the logistics of his *Principia Mathema-*

---

[1] Bertrand Russell, *My Philosophical Development*, (London: George Allen & Unwin, 1959), p. 40.

[2] Ibid., p. 11.

tica. After this shift Russell had absolutely no patience for the kinds of conceptual issues with which he had struggled in the final decade of the century.

The English tradition which had nurtured the young Russell was also displaced soon after he published his *Essay.* The change was not as cataclysmic as Russell's personal one, but it was equally fundamental. In the first decades of the new century, the unitary view of truth, which had for so long provided the justificatory framework for geometrical study, crumbled into a more fragmented and specialized view of knowledge and learning. This disintegration allowed mathematicians to develop their subject without integrating it into an essentially scientific unitary view of truth. This, in turn, made it possible to relax the descriptive focus which had so long formed the justification for English mathematical work.

The matriculation requirements at Cambridge and Oxford can be viewed as institutional indicators of the views of knowledge being supported within their culture. This is a highly conservative gauge, and major reforms were effected here only after they had been long advocated from outside of the universities. John Perry likened the educational reforms of the first decade of the twentieth century to the collapse of the walls of Jericho; it might appear to have been precipitous but required years of circling to effect.[3] Reform may have been slow to come, but when it did, it heralded a radical shift in the English view of mathematical study. Nineteenth-century mathematics had been supported and defined within the integrated notion of a liberal education. As that notion was adjusted to meet the demands of a changing world, the English view of what constituted mathematics changed also.

The spirit behind the movement for curricular reform at the university level was essentially the same as that which had grown up at the secondary level. A large part of the criticism that fueled the forces for change was directed against the structure of an education that pointed towards performance on a single examination. By the end of the century, that structure was undeniably odd. In order to do creditably on the Tripos, one needed to engage a coach, who would both teach the material which was likely to be tested and train

---

[3] John Perry, "Mathematics in the Cambridge Locals," *Nature,* 67 (1903):82.

the students in examination strategies. Often the latter concern seemed to overshadow the former, and the twists and turns of the Tripos, as opposed to relevant mathematics, became the focus of the students' attention.

This problem was exacerbated by the way the examination was marked. The results of the examination, which could be critically important to a student's future, were grouped into three major categories—Wrangler, First Class and Second Class—and students were rank listed within each category. The necessity of making distinctions among students which would be precise and fine enough to rank each of them individually meant that the examination was full of picky details, trick questions and memorized proofs. Thus students had to devote considerable time to exploring the obscure nooks and crannies of mathematical detail in order to prepare themselves for the Tripos competition. Furthermore, the examination contained more problems than anyone could possibly finish in the time allotted. This design was intended to give students choice. However, in practice, it meant that for the good students, the speed with which they wrote became a major factor determining how well they placed. For this group, the examination could be as much a handwriting race as a test of learning. It is little wonder that many who learned mathematics within this system complained bitterly about it afterwards.

At the end of the century, a number of reforms were proposed to deal with the educationally destructive idiosyncracies of the Tripos. The first, instituted in 1892, allowed students to take the Tripos in the course of the second rather than the third year of their career. This was intended to free them to pursue their own interests in the final two years. In the event, however, this attempt to liberate students from the excessive rigors of Tripos preparation was ineffective. The rankings were simply too important for students to jeopardize their standing with inadequate preparation.

Therefore an additional reform was proposed which was designed to mitigate the intense pressure to do well on the examination. This reform, seriously suggested in 1899 but not instituted until 1907, involved abolishing the order of merit; instead of establishing and publishing a strict rank ordering of all who had passed the examination, the results would be grouped alphabetically in three categories: Wrangler, Senior Optime and Junior Optime. By eliminating the need to make fine distinctions among candidates, the reform was designed to eliminate many of the excesses which abounded on the late-century Tripos: the picky questions, trick problems and extensive memory work which had come to dominate the examination. It led to a looser Tripos, requiring less specific knowledge and allowing more freedom in mathematical approach. In this way it removed the incentive to over-prepare for the examination and effectively freed students to take it in their second year.

Implicit in these Tripos reforms were major changes in the English view of the nature of mathematics itself. One aspect of the change lies in the scope of mathematics. The mounting wail of frustration against the crabbed excesses of Tripos detail implied a fundamental critique of the nineteenth-century view of the nature of the subject. Cayley's 1883 description of mathematics suggests the kind of orientation that led him to fiercely defend the old Tripos until his death.

> It is difficult to give an idea of the vast extent of modern mathematics. This word 'extent' is not the right one: I mean extent crowded with beautiful detail—not an extent of mere uniformity such as an objectless plain, but of a tract of beautiful country seen at first in the distance, but which will bear to be rambled through and studied in every detail of a hillside and valley, stream, rock, wood, and flower.[4]

For Cayley, much of the joy and value of mathematical knowledge lay in its detail. The activities the reformers damned as cramming for the Tripos, he would have defended as truly learning mathematics.

At a more basic level, the reformers wanted to change the goals of the Cambridge mathematical education. Their attempt to remove the wealth of detail from the student's vision in order to move the Tripos into his second year involved a major move away from the mid-century vision of the liberal education. It meant that the Tripos became a preliminary hurdle rather than the primary goal of the Cambridge education.

---

[4] Arthur Cayley, "Presidential Address," *Report of the Fifty-third Meeting of the BAAS held at Southport in September 1883*, (London: John Murray, 1884), p. 25.

This shift meant that mathematics was perceived differently than it had been before. In the mid-century view, the elementary mathematical work necessary for Part I of the Tripos was an end in itself. It encompassed a kind of knowledge useful for any person in any place under any circumstances. It is worth quoting at some length from Isaac Todhunter's defense of the nineteenth-century system to see the parameters of this way of thinking.

> Many persons seem to hold that the great function of an examination is the supply of a continued stream of eminent mathematicians. . . . Such words as the following are based on this notion: 'Merely to be able to integrate, to solve differential equations, to work the hardest of the Senate House Problems &c. &c. is *not* to be a mathematician. To deserve the name a man must have some of the creative faculty . . . if ever so little.' I venture to disagree with what is here said and implied. I take the list of accomplishments which is recorded only to be depreciated, and . . . I say that the man who possesses these *is* a mathematician. He may never have the leisure which official dignity and emoluments ensure, he may be shut out from every academical advantage on account of his father's creed, he may be compelled to occupy his time in constant drudgery for the sake of subsistence, or he may devote his ability to objects of deeper than scientific interest; for all these reasons he may contribute little to the advancement of the subjects which he is qualified to handle, but he is potentially a mathematician.[5]

For Todhunter, being a mathematician was a condition which was not necessarily connected with mathematical productivity. It was an almost spiritual accomplishment which, once achieved, could not be taken away.

In the early twentieth-century reformers' view, on the other hand, learning the material on the Tripos was a means to some other end. To quote a late-century advocate of reform in *Nature:*

> By taking the Tripos in their second year, men who intend to study subjects such as physics or engineering will be able to gain a preliminary knowledge of mathematics, with indications as to how to extend

---

[5] I[saac] Todhunter, *The Conflict of Studies and Other Essays on Subjects Connected with Education,* (London: Macmillan and Co., 1873), p. 212.

their knowledge in any special branch which they may need in their future course.[6]

From this perspective, the Tripos was part of a general as opposed to a liberal education. It was intended to show the students the range of mathematics available for use in a variety of fields, rather than forming their adult character.

As the above quotation from *Nature* suggests, the arguments for taking the Tripos early were often couched in terms of its advantage to those who wished to use mathematics in pursuit of other special areas. The practical needs of engineers or physicists were a major focus of reform interest. This reflects the ever-increasing strength of the middle-class professional interests which were first expressed in the 1870s. It also reflects the particular situation at the turn of the century. At that time England was gripped in a xenophobic reaction to Germany's growing strength. Their fear that England was falling behind as an international power was fed by the administrative and military failures of the Boer War. A large part of their concern focused on the state of science and scientific education in England.[7]

One of the clearest expressions of concern about the state of English science is Norman Lockyer's Presidential Address to the BAAS in 1903. He entitled the opening section "The Influence of Brain-power on History." In it Lockyer argued that the strength of a nation depended as much on its mental as its material resources. He berated the English nation because "we have lacked the strengthening of the national life produced by fostering the scientific spirit among all classes and along all lines of the nation's activity; . . . we have not learned that it is the duty of a State to organize its forces as carefully for peace as for war; that Universities and other teaching centres are as important as battleships or big battalions."[8] In the insecure atmosphere of prewar England, this kind of argument was frighteningly cogent.

---

[6]"The Proposed Changes in the Mathematical Tripos," *Nature,* 61 (1899):107.

[7]For a fuller discussion of these issues, see George Haines IV, *Essays on German Influence upon English Education and Science, 1850–1919,* Connecticut College Monograph no. 9, (Connecticut College in association with Archon Books, 1969).

[8]Norman Lockyer, "Presidential Address," *Report of the Seventy-third Meeting of the BAAS held at Southport in September 1903,* (London: John Murray, 1904), pp. 5–6.

Lockyer and others like him did not just advocate more scientific education; they advocated a reform of the institutions which were already in place. The German educational model was particularly attractive in their reform vision. Its salient characteristics can be drawn from the following description from an article in *Nature*.

> The universities in Germany owe their hold over the intellectual life of the people to their unreserved acceptation of the scientific spirit, that is to say, the spirit of inquiry and free investigation into all the departments of learning. The university is defended and vindicated . . . primarily as an institution for research and the advancement of knowledge, and secondarily as a place of education; secondarily, not from the mistaken notion that education is considered less important than the expansion of the limits of knowledge, . . . but because the most important part of a university education is considered to be the actual contact with the fountains of knowledge and the acquisition of a capacity to grapple with original sources and to form an independent opinion.[9]

A key aspect of the new vision many reformers were trying to establish lay in their support of research. Lockyer referred to research as "the most powerful engine of education that we possess."[10] National strength and progress was seen to be directly linked to a kind of disinterested research which the reformers claimed had been too long neglected in England.

Clearly most of xenophobically generated reforming zeal was directed towards creating an obviously productive group of practically educated mathematicians. A corollary to these arguments, however, had the effect of justifying the pursuit of totally disinterested, pure mathematics. Lockyer followed the passage quoted above with considerations of the benefits of the proposed reforms for those who wished to develop pure mathematics. The same spirit of research support permeated many of the justifications for reform. Thus, *Nature* supported the second-year Tripos in the following terms:

> The second year Tripos will be of advantage also to the better mathematical men, who now spend half of their third year in revision and

---

[9] G. S., "Higher Education in Germany," *Nature*, 75 (1907):338.
[10] Lockyer, "Address," (1904), p. 22.

in acquiring facility of solving artificial problems. Under the proposed regulations these men will have two years after Part I (instead
of one) in which to become acquainted with the ideas and methods
of modern mathematics. This will be of special advantage to men who
intend to devote themselves to mathematical research.[11]

A new wind blows through this view of mathematics and the image
it presents of the mission of the best mathematicians. The subject is
not presented as primarily defined and justified by its mind-forming
educational function. Todhunter's vision of the mathematician as a
beautifully formed human being is replaced by a professional image:
it is assumed that the best mathematicians will go into research.

Emphasizing the central importance of disinterested mathematical research represented a major and wrenching change in English
mathematical practice at the turn of the century. Russell's career in
the 1890s hints at the change to come. After he finished Cambridge,
Russell rejected its confinement and went to Germany. There he
was introduced to a totally different kind of mathematics, notably
that of Felix Klein and his school. One way of seeing the difference
is philosophical; the continental approach to mathematics was significantly more formal than that to be found in the England of Russell's origin. Another difference, however, was social and related to
the organization of the intellectual community. Klein was a research
mathematician who self-consciously drew a community of like-
minded people around him, people whose primary interest was
developing new mathematical results.[12] This was radically different
from the English system, where educational issues were the focus
for most of the university community.

Russell's response to these novel influences was rather timid.
After a brief stay, he returned to England and to Cambridge. His
first book shows that this was an intellectual as well as a physical
homecoming; the conceptual treatment of projective geometry on
which his *Essay* turns reflects the English educational focus more
than it does the more abstract German view.

[11] "Proposed Changes," (1899), p. 107.
[12] There is a considerable literature about Felix Klein and his career which is presented in
David Rowe's review of Renate Tobies, *Felix Klein,* Karl-Heinz Manegold, *Universität, Techn-
ische Hochschule und Industrie* and Lewis Pyenson, *Neohumanism and the Persistence of Pure Mathematics in Wilhelmian Germany,* in *Historia Mathematica,* 12 (1985):278–91.

A more dramatic response to the exotic German views of the primary role of the mathematician can be found in the career of William H. Young, who coached at Cambridge when Russell was there as a student. Young's early career exemplifies the nineteenth-century educational tradition in English mathematics. He studied at Peterhouse, Cambridge, from 1881 to 1885 and, from 1890 to 1897, worked as a mathematics coach. During this period he did no original mathematical work at all, being content to pursue the study entirely in an educational context. In June of 1896, however, Young married Grace Chisholm, who had gone from her Cambridge degree in 1893 to study mathematics in Göttingen under Felix Klein. In 1895, Chisholm had submitted a dissertation which earned her the first German doctorate ever granted to a woman. In 1897, soon after the birth of their first child, the Youngs left Cambridge, eventually settling in Göttingen from 1900 to 1908.

The Youngs' move away from England, which never again served as their permanent residence, was not just geographical. It was a statement about the way a mathematician should spend his life. Upon leaving England, they began to pursue original mathematical research. William Young published his first paper in 1898 when he was thirty-six; this was the beginning of a research career which resulted in the publication of more than two hundred papers before his death in 1942.[13] Young's life can serve as a demonstration of the impact of the German research ideal on English mathematical life at the turn of the century. His midlife upheaval illustrates the radical nature of the adjustment which was involved in moving from the English educational tradition in mathematics to the German research one.

The change in the way mathematicians pursued their subject, which accompanied the turn of the century educational reforms,

---

[13] The best source of information on the Youngs is Ivar Grattan-Guiness, "University Mathematics at the Turn of the Century: Unpublished Recollections by W. H. Young," *Annals of Science*, 28 (1973):369–84; Ivar Grattan-Guiness, "A Mathematical Union: William Henry and Grace Chisholm Young," *Annals of Science*, 29 (1972):105–86. Recently there has been increased interest in understanding the dynamics of the Youngs' marriage and professional relationship. The importance of Grace to Will's research success has been emphasized in Sylvia Wiegand, "Grace Chisholm Young," *Association for Women in Mathematics Newsletter*, 7 (1977):5–10.

carried implications which went to the very heart of the subject itself. Embracing the research ideal freed England's mathematicians to unabashedly pursue mathematics "for its own sake" without needing the justificatory educational structure in which it had been embedded for so long. From this perspective one could view the subject as wholly abstract without being damning in the way Huxley had been. The subject so described might not appeal to all people, but within a research ideology it could nonetheless be deemed a worthy pursuit. This approach is lucidly expressed in G. H. Hardy's classic, *A Mathematician's Apology,* first published in 1941.

Hardy, who placed first on Part II of the unreformed Tripos of 1900, was the embodiment of the new breed of twentieth-century Cambridge mathematician. He opened his musings with the following self-deprecatory but revealing comment:

> It is a melancholy experience for a professional mathematician to find himself writing about mathematics. The function of a mathematician is to do something, to prove new theorems, to add to mathematics, and not to talk about what he or other mathematicians have done. Statesmen despise publicists, painters despise art-critics, and physiologists, physicists, or mathematicians have usually similar feelings; there is no scorn more profound, or on the whole more justificable, than that of the men who make for the men who explain. Exposition, criticism, appreciation, is work for second-rate minds.[14]

These opening words are an unequivocal damnation of the value of seriously approaching mathematics in an educational context. In its stead, Hardy offered a dynamic vision in which the mathematician's role was to push forward the frontiers of mathematical knowledge.

Having taken his anti-educational stand, Hardy was faced with the question of what positive value mathematical study had. His answer was firmly nonutilitarian; he denied that mathematics had any practical value at all. Instead, he defended the study of mathematics as a pastime, eminently suited for a certain kind of intelligence. His "apology" was for a life-choice which, he argued, at least had the rare value of being harmless.

---

[14] G. H. Hardy, *A Mathematician's Apology,* (Cambridge: University Press, 1982), p. 61.

Hardy's characterization of the essence of the mathematical career provides a stark contrast to the nineteenth-century view of the mathematical life. Research was an important aspect of the lives of men like De Morgan, Clifford, Cayley, Sylvester and their contemporaries, but it did not dominate them. Their enthusiastic explorations of the meaning of their subject certainly did not mirror the melancholy Hardy felt as he did the same. The narrowly focused research ideal Hardy had pursued in his professional life was radically different from the multifarious lifestyles of his nineteenth-century predecessors.

The difference in these professional profiles is critically important to interpretations of the foundations of mathematics itself. In Chapter 4, it was emphasized that all discussions of what mathematics is, whether they are pursued in the context of elementary education or advanced philosophy, are in some sense foundational questions. This point can be made in a somewhat different way as well. Mathematics is defined in an important way by what mathematicians do. The importance of this factor in the formulation of foundational ideas about what mathematics is, is illustrated by the early twentieth-century situation in England.

Throughout the second half of the nineteenth century, English characterizations of mathematics were persistently conceptual. This interpretation was in line with the educational and social function of the mathematics they pursued. This changed rather abruptly in the early years of the twentieth century, however. With the collapse of the educational walls of Jericho which had so long protected the conceptual view of mathematical truth, that view fell as well. The basic outlines of this new perspective were clearly laid out by Alfred North Whitehead in a section entitled "Axioms of Geometry" within a longer article, "Geometry," for the eleventh edition of the *Encyclopaedia Britannica*. Here, Whitehead proposed the following "Definition of Abstract Geometry":

> Geometry has been defined as 'the study of series of two or more dimensions.' It has also been defined as 'the science of cross classification.' These definitions are founded upon the actual practice of mathematicians in respect to their use of the term 'Geometry.' Either of them brings out the fact that geometry is not a science with a

determinate subject matter. It is concerned with any subject matter
to which the formal axioms may apply.[15]

The conceptual approach to geometry simply had no place in Whitehead's view. As he wrote elsewhere, "The whole apparatus of special
indefinable mathematical concepts, and special *a priori* mathematical premises, respecting number, quantity and space, has
vanished."[16]

The radical shift in perspective Whitehead's words reflect was
made possible by the changes in the professional roles of mathematicians signaled by the collapse of the nineteenth-century Tripos.
Removing the study of mathematics from its position in the liberal
education allowed the focus of what was mathematically acceptable
or adequate to shift away from questions about the nature of knowledge. It meant specifically that mathematicians could assess validity
in their own way, without justifying their views in terms of models
of truth generated in the objective sciences. Under these conditions,
the nature of the subject matter of mathematics need not be a pressing problem. If mathematics were not a science, it need not even
have a subject matter. This approach meant that all of the questions
about the relation of geometry to space, which had so long preoc-
cupied nineteenth-century Englishmen, could be simply dropped.
Geometry could become a formal study, untrammeled by worries
about the spatial meanings of its definitions or results. Whitehead's
Encyclopaedia article points to the sociological basis for the nonconceptual view of geometry. "These definitions are founded upon the
actual practice of mathematicians in respect to their use of the term
'Geometry,'" he wrote. The new view of mathematical foundations
he presented reflected the views of a new, twentieth-century mathematical community. The function or mission of this group was
redefined in such a way that they could legitimately pursue mathematics without regard to questions about epistemology.

---

[15] *Encyclopaedia Britannica*, 11th ed. (1910–11), "The Axioms of Geometry," s.v. "Geometry,"
by Alfred North Whitehead.

[16] Alfred North Whitehead, "Presidential Address: Mathematics and Physics Section," *Report
of the Eighty-sixth Meeting of the BAAS held at Newcastle in August 1916*, (London: John Murray,
1917), p. 361.

The issues remained, however; they were just defined as the concern of different professional groups. This Whitehead recognized, and he opened his article with a consideration of the kinds of controversies which in the preceding century had been generated by non-Euclidean geometries. In the preamble to his discussion of abstract geometry, Whitehead focused on the kinds of questions about how humans know space that had for so long shaped English geometrical practice. Whitehead gave these issues short shrift, however. To enter into considerations about whether space was known *a priori* or through experience would, he explained, "merge into a general treatise on epistemology." Another angle on the question, "the derivation of our perception of existent space from our various types of sensation," he dismissed as "irrelevant to this article," it being "a question for psychology."[17] With these remarks, Whitehead defined the questions which had for so long been clustered around mathematical foundations as outside of the legitimate purview of mathematics. He thus indicated that it was acceptable for mathematicians to ignore such questions in their work.

Whitehead's treatment reveals some of the parameters of a new organization of knowledge which early in the twentieth century superseded the nineteenth-century unitary view of truth. It was a fragmented view which allowed mathematicians to pursue their study in ways which cut it off from all of the other sciences. It sharply separated the concerns of epistemology from those of psychology, and both of these from mathematics. It represented a radical break from the nineteenth-century tradition epitomized by Whewell or Mill, which tried to bring all of knowledge under a single philosophical umbrella.

The reorganization of knowledge which Whitehead articulated at the turn of the century had a profound impact on the professional lives of many of England's intellectuals. Whitehead, for example, began his career as a nineteenth-century mathematician and ended it as a twentieth-century philosopher. Others, like Young or Hardy, gladly abandoned the nineteenth-century vision of epistemological integration and enthusiastically embraced their new, specialized role

---

[17] Whitehead, "Axioms," (1910–11):703.

as pure mathematicians. Others, like Perry, stayed in education; and Russell followed his own idiosyncratic trajectory.

Whatever the individual responses, the important point is that at the turn of the century England's mathematicians faced a different world than they had before. With the collapse of the unitary worldview which had supported the nineteenth-century Tripos for so long came a new mathematical vision. The establishment of a recognized and viable mathematical research community in England opened the door for a formal view of the subject. This view had not been viable within the nineteenth-century perspective which kept mathematics essentially educational and insisted that its essence mirror all forms of human knowledge. It only became possible when the the social and cultural role of the subject had been radically redefined. With this change, the nineteenth-century tradition abruptly disappeared. Books like Russell's *Essay* were quickly forgotten. Geometry ceased to be the quintessential science, and England's mathematicians began to pursue a new role and a new vision into the twentieth century.

References

[Abbott, Edwin Abbott]. A Square. *Flatland; A Romance of Many Dimensions.* London: Seeley, 1884.

Agassi, Joseph. "Sir John Herschel's Philosophy of Success." *Historical Studies in the Physical Sciences* 1 (1969):1–36.

Airy, George Biddell. *Autobiography of Sir George Biddell Airy.* Edited by Wilfred Airy. Cambridge: University Press, 1896.

Arnold, Matthew. *Higher Schools and Universities in Germany.* 2nd. ed. London: Macmillan and Co., 1892.

Ashby, Eric. *Technology and the Academics: An Essay on Universities and the Scientific Revolution.* New York: St. Martin's Press, 1958.

Association for the Improvement of Geometrical Teaching. *Annual Report.* 19 vols., 1871–1893.

Association for the Improvement of Geometrical Teaching. *Syllabus of Plane Geometry.* London: Macmillan and Co., 1875.

Babbage, Charles. *The Exposition of 1851; or, Views of the Industry, the Science, and the Government of England.* London: J. Murray, 1851. Reprint. Farnborough, Hants, England: Gregg International Publishers, 1969.

Babbage, Charles. *The Ninth Bridgewater Treatise. A Fragment.* London: J. Murray, 1837.

Babbage, Charles. *Passages from the Life of a Philosopher.* London: Longman, Green, Longman, Roberts & Green, 1864. Reprint. Farnborough, Hants, England: Gregg International Publishers, 1969.

Babbage, Charles. *Reflections on the Decline of Science in England and on Some of its Causes.* London: Printed for B. Fellowes, 1864. Reprint. Farnborough, Hants, England: Gregg International Publishers, 1969.

Baker, J. T. "An Historical and Critical Examination of English Space and Time Theories from Henry Moore to Bishop Berkeley." Ph.D. Dissertation, Sarah Lawrence College, 1930.

245

Ball, Robert S[tawell]. "A Dynamical Parable: Presidential Address to the Mathematical and Physical Section." *Report of the Fifty-seventh Meeting of the BAAS held at Manchester in August and September, 1887.* London: John Murray, 1888. 569–79.

Ball, Robert S. "The Non-Euclidean Geometry." *Hermathena* 3 (1877–79): 500–41.

Ball, Robert S. "On the Theory of the Content." (Read Dec. 12, 1887). *The Transactions of the Royal Irish Academy* 29 (1887–92): 123–82.

Ball, Walter William Rouse. *A History of the Study of Mathematics at Cambridge.* Cambridge: University Press, 1889.

Ball, Walter William Rouse. *A Short Account of the History of Mathematics.* London: Macmillan and Co., 1888.

Barnes, Barry. *Interest and the Growth of Knowledge.* Boston: Routledge & K. Paul, 1977.

Barnes, Barry and Stephen Shapin, ed. *Natural Order: Historical Studies of Scientific Culture.* Beverly Hills: Sage Publications, 1979.

Becher, Harvey W. "William Whewell and Cambridge Mathematics." *Historical Studies in the Physical Sciences* 11 (1980): 1–48.

Bell, Eric Temple. *The Development of Mathematics.* 2nd ed. New York: McGraw-Hill, 1945.

Bellot, Hugh Hale. *University College London, 1826–1926.* London: University of London Press, 1929.

Beltrami, E. "Essai d'interpretation de le géométrie non-Euclidienne." Translated by M. J. Houël. *Annales scientifiques de l'Ecole Normale Supérieur* 6 (1869): 251–88.

Beltrami, E. "Saggio di interpretazione della geometria non-euclidea." *Giornale di matematiche* 6 (1868): 284–312.

Besant, W. H. *The Student's Guide to the University of Cambridge.* 5th ed. Cambridge: Deighton, Bell and Co., 1893.

Blake, Ralph Mason, Curt J. Ducasse and Edward H. Madden. *Theories of Scientific Method: The Renaissance through the Nineteenth Century.* Edited by Edward H. Madden. Seattle: University of Washington Press, 1960.

Bloor, David. *Knowledge and Social Imagery.* Boston: Routledge & Kegan Paul, 1976.

Bonola, Roberto. *Non-Euclidean Geometry; A Critical and Historical Study of its Development.* Authorized translation by H. S. Carslaw. Introduction by Frederigo Enriques. With a Supplement containing the George Bruce Halsted translation of "The Science of Absolute Space," by John Bolyai, and "The Theory of Parallels," by Nicholas Lobachevski. New York: Dover Publications, 1955.

Boole, George. *An Investigation of the Laws of Thought on Which are Founded the Mathematical Theories of Logic and Probabilities.* London: Walton and Maberly, 1854. Reprint. New York: Dover Publications, 1951.

Bourbaki, Nicolas. *Elements d'histoire des mathématiques.* 2d. éd. rev., corrigée, augmentée. Paris: Hermann, 1969.

Bowne, G. D. *Philosophy of Logic, 1880–1908.* London: Mouton & Co., 1966.

Boyer, Carl Benjamin. *History of Analytic Geometry*. New York: Scripta Mathematica, 1956.

Boyer, Carl Benjamin. *A History of Mathematics*. New York: Wiley, 1968.

Bradley, F. H. *The Principles of Logic*. 2 vols. London: Oxford University Press, 1922.

Bradley, James. "Hegel in Britain: A Brief History of British Commentary and Attitudes." *The Heythrop Journal* 20 (1979): 1–24,163–82.

Bristed, Charles Astor. *Five Years in an English University*. 2 vols. 3rd ed. New York: G. P. Putnam and Sons, 1873.

Brock, W. H. "Geometry and the Universities: Euclid and his Modern Rivals, 1860–1900." *History of Education* 4 (1975): 21–35.

Brock, William H. and Roy M. MacLeod, eds. *Natural Knowledge in Social Context: The Journals of Thomas Archer Hirst, F.R.S.* London: Mansell, 1980.

Brown, Alan Willard. *The Metaphysical Society: Victorian Minds in Crisis, 1869–1880*. New York: Columbia University Press, 1947.

Buttman, Gunther. *The Shadow of the Telescope: A Biography of John Herschel*. Translated by B. E. J. Pagel. Edited with an Introduction by David S. Evans. New York: Charles Scribner's Sons, 1970.

Butts, Robert E., ed. *William Whewell's Theory of Scientific Method*. Pittsburgh: University of Pittsburgh Press, 1968.

Cajori, Florian. *A History of Mathematics*. New York: Macmillan Company, 1922.

Cajori, Florian. *Mathematics in Liberal Education: A Critical Examination of the Judgments of Prominent Men of the Ages*. Boston: Christopher Publishing House, 1928.

Campbell, Lewis and William Garnett. *The Life of James Clerk Maxwell*. London: Macmillan and Co., 1882.

[Cannon, Susan Faye]. Walter Cannon. "John Herschel and the Idea of Science." *Journal of the History of Ideas* 22 (1961): 215–39.

Cannon, Susan Faye. *Science in Culture: The Early Victorian Period*. New York: Dawson and Science History Publications, 1978.

Cardwell, Donald Stephen Lowell. *The Organisation of Science in England*. Revised ed. London: Heinemann Educational, 1972.

Cayley, Arthur. *The Collected Mathematical Papers of Arthur Cayley*. 11 vols. Cambridge: University Press, 1889–97.

Cayley, Arthur. "A Memoir on Abstract Geometry." (Read Dec. 16, 1869). *Philosophical Transactions on the Royal Society* 160 (1870): 51–63.

Cayley, Arthur. "Non-Euclidean Geometry." (Read Jan. 27, 1890). *Transactions of the Cambridge Philosophical Society* 15 (1894): 36–61.

Cayley, Arthur. "Note on Lobatchewsky's Imaginary Geometry." *Philosophical Magazine* 4th series. 29 (1865): 231–33.

Cayley, Arthur. "On the Non-Euclidean Geometry." *Mathematische Annalen* 5 (1872): 630–34.

Cayley, Arthur. "On the Non-Euclidean Plane Geometry." *Proceedings of the Royal Society of London* 37 (1884): 82–102.

Cayley, Arthur. "Presidential Address." *Report of the Fifty-third Meeting of the BAAS held at Southport in September 1883*. London: John Murray, 1884. 3–37.

Cayley, Arthur. "A Sixth Memoir upon Quantics." *Philosophical Transactions of the Royal Society* 149 (1859): 61–90.

Chasles, Michel. *Aperçu historique sur l'origine et le développement des méthodes en géométrie, particulièrement de celles qui se rapportent à la géométrie moderne.* Bruxelles: M. Hayez, 1837.

Chrystal, George. "Address to the Mathematics and Physics Section." *Report of the Fifty-fifth Meeting of the BAAS held at Aberdeen in September 1885.* London: John Murray, 1886. 889–96.

Chrystal, George. "Non-Euclidean Geometry." *Proceedings of the Royal Society of Edinburgh* 10 (1879–80): 638–65.

Chrystal, George. "s. v. Mathematics." *Encyclopaedia Britannica.* 9th ed. 1883.

Clifford, William Kingdon. *The Common Sense of the Exact Sciences.* Edited and with a preface by Karl Pearson. Newly edited and with an introduction by James R. Newman. Preface by Bertrand Russell. New York: A. A. Knopf, 1946.

Clifford, William Kingdon. *Lectures and Essays by the late William Kingdon Clifford.* Edited by Leslie Stephen and Sir Frederick Pollock. New York: Macmillan Company, 1901.

Clifford, William Kingdon. *Mathematical Papers.* Edited by Robert Tucker. With an introduction by H. J. Stephen Smith. London: Macmillan and Co., 1882.

Clifford, William Kingdon. "On the Aims and Instruments of Scientific Thought." *Macmillan's Magazine* 26 (1872): 499–512.

Clifford, William Kingdon. "On the Space-Theory of Matter." (Read Feb. 21, 1870). *Proceedings of the Cambridge Philosophical Society* 2 (1864–1876): 157–58.

Clifford, William Kingdon. "The Philosophy of the Pure Sciences, Pt. 1., The Statement of the Question." *Contemporary Review* 24 (1874): 712–27.

Clifford, William Kingdon. "The Philosophy of the Pure Sciences, Pt. 3., The Universal Statements of Arithmetic." *The Nineteenth Century* 5 (1879); 513–22.

Clifford, William Kingdon. "The Philosophy of the Pure Sciences, Pt. 2., The Postulates of the Science of Space." *Contemporary Review* 25 (1875): 360–76.

Clifford, William Kingdon. "The Unreasonable." *Nature* 7 (1873): 282.

Clock, Daniel Arwin. "A New British Concept of Algebra." Ph.D. Dissertation, University of Wisconsin, 1964.

Comte, Auguste. *The Philosophy of Mathematics.* Translated from *Cours de philosophie positive* by W. M. Gillespie. New York: Harper & Brothers, 1851.

Coolidge, Julian Lowell. *A History of Geometrical Methods.* Oxford: Clarendon Press, 1940. Reprint. New York: Dover Publications, 1963.

Coolidge, Julian Lowell. "The Rise and Fall of Projective Geometry." *American Mathematical Monthly* 41 (1934): 217–28.

Corsi, Pietro. *Science and Religion: Baden Powell and the Anglican Debate, 1800–1860.* Cambridge: University Press, 1987.

Courant, R[ichard]. "Bernhard Riemann und die Mathematik der letzen hundert Jahre." *Die Naturwissenschaften* 36 (1926); 813–18, 1265–77.

Couturat, Louis. "Essai sur les fondements de la géométrie par Bertrand Russell." *Révue de métaphysique et de morale* 6 (1898): 354–80.

Cox, Homersham. "Homogeneous Co-ordinates in Imaginary Geometry and their Application to Systems of Forces." *The Quarterly Journal of Pure and Applied Mathematics* 18 (1882): 178–215.

Coxeter, Harold Scott Macdonald. *Introduction to Geometry.* 2nd ed. New York: Wiley, 1969.

Coxeter, Harold Scott Macdonald. *Non-Euclidean Geometry.* Toronto: University of Toronto Press, 1942.

Crosland, Maurice and Crosbie Smith. "The Transmission of Physics from France to Britain: 1800–1840." *Historical Studies in the Physical Sciences* 9 (1978): 1–61.

Crowe, Michael J. *A History of Vector Analysis: The Evolution of the Idea of a Vectorial System.* Notre Dame: University of Notre Dame Press, 1967.

Daniels, Norman. "Thomas Reid's Discovery of a Non-Euclidean Geometry." *Philosophy of Science* 39 (1972): 219–34.

Daniels, Norman. *Thomas Reid's* Inquiry: *The Geometry of Visibles and the Case for Realism.* Forward by Hilary Putnam. New York: Burt Franklin & Co., 1974.

Daston, Lorraine J. "The Physicalist Tradition in Early Nineteenth Century French Geometry." *Studies in the History and Philosophy of Science* 17 (1986): 269–95.

Davie, George Elder. *The Democratic Intellect: Scotland and her Universities in the Nineteenth Century.* Edinburgh: University Press, 1961.

De Morgan, Augustus. "On Divergent Series, and Various Points of Analysis Connected with Them." (Read Mar. 4, 1844). *Transactions of the Cambridge Philosophical Society* 8 (1849): 182–203.

De Morgan, Augustus. "On the Foundation of Algebra." (Read Dec. 9, 1839). *Transactions of the Cambridge Philosophical Society* 7 (1842): 173–87.

De Morgan, Augustus. "On the Foundation of Algebra, No. II." (Read Nov. 29, 1841). *Transactions of the Cambridge Philosophical Society* 7 (1842): 287–300.

De Morgan, Augustus. "On the Foundation of Algebra, No. III." (Read Nov. 27, 1843). *Transactions of the Cambridge Philosophical Society* 8 (1849); 141–42.

De Morgan, Augustus. "On the Foundation of Algebra, No. IV., On Triple Algebra." (Read Oct. 28, 1844). *Transactions of the Cambridge Philosophical Society* 8 (1849): 241–54.

De Morgan, Augustus. "On the Signs + and − in Geometry (continued), and On the Interpretation of the Equation of a Curve." *The Cambridge and Dublin Mathematical Journal* 7 (1852): 242–49.

De Morgan, Augustus. *On the Study and Difficulties of Mathematics.* 3rd ed. Chicago: Open Court Publishing Co., 1910.

De Morgan, Augustus. "Response to Wilson's Reply." *The Athenaeum* 2130 (1868): 241–42.

De Morgan, Augustus. "Review of *Geometry without Axioms.*" *Quarterly Journal of Education* 7 (1834): 105–15.

De Morgan, Augustus. "Review of George Peacock, *A Treatise on Algebra.*" *Quarterly Journal of Education* 9 (1835): 91–110, 293–311.

De Morgan, Augustus. "Review of Whewell's *Novum Organon Renovatum.*" *The Athenaeum* 1628 (1859): 42–44.

De Morgan, Augustus. "Review of Whewell's *Philosophy of Discovery.*" *The Athenaeum* 1694 (1860): 501–503.

De Morgan, Augustus. "Review of Wilson, *Elementary Geometry.*" *The Athenaeum* 2125 (1868): 71–73.

De Morgan, Augustus. "Short Supplementary on the First Six Books of Euclid's Elements." *The Companion to the Almanac.* London: Charles Knight, 1849. 5–20.

De Morgan, Augustus. "Speech of Professor De Morgan." *Proceedings of the London Mathematical Society* 1 (1865–66): 1–9.

De Morgan, Augustus. *Trigonometry and Double Algebra.* London: Taylor, Walton and Maberly, 1849.

De Morgan, Sophia Elizabeth. *Memoir of Augustus De Morgan.* London: Longmans, Green and Co., 1882.

Dixon, Edward T[ravers]. *An Essay on Reasoning.* Cambridge: Deighton, Bell and Co., 1891.

Dixon, Edward T. *The Foundations of Geometry.* Cambridge: Deighton, Bell and Co., 1891.

Dixon, Edward T. "Letter." *Nature* 47 (1892): 127.

Dixon Edward T. "On the Distinction between Real and Verbal Propositions." *Mind* n.s. 2 (1893): 339–46.

Dixon, Edward T. *A Paper on the Foundations of Projective Geometry.* Cambridge: Deighton, Bell & Co., 1898.

Dodgson, Charles L. *Euclid and his Modern Rivals.* London: Macmillan and Co., 1885.

Douglas, Mrs. Stair [Janet Mary Douglas], ed. *The Life and Selections from the Correspondence of William Whewell.* London: C. K. Paul & Co., 1881.

Dubbey, J. M. "The Introduction of the Differential Notation to Great Britain." *Annals of Science* 19 (1963): 37–48.

Dubbey, J. N. *The Mathematical Work of Charles Babbage.* Cambridge: University Press, 1978.

Ellegaard, Alvar. *Darwin and the General Reader: The Reception of Darwin's Theory of Evolution in the British Periodical Press, 1859–1872.* Göteborg: Gothenburg Studies in English, no. 8., 1958.

Ellegaard, Alvar. *The Readership of the Periodical Press in Mid-Victorian Britain.* Göteborg: Göteborgs Universitets Arsskrift, 1957.

Ellis, Robert Leslie. *The Mathematical and Other Writings of Robert Leslie Ellis.* Edited by William Walton. With a Biographical Memoir by the Very Reverend Harvey Goodwin. Cambridge: Deighton, Bell and Co., 1863.

Enros, Philip. "The Analytical Society: Mathematics at Cambridge University in the Early Nineteenth Century." Ph.D. Dissertation, University of Toronto, 1979.

Enros, Philip C. "The Analytical Society (1812–1813): Precursor of the Renewal of Cambridge Mathematics." *Historia Mathematica* 10 (1983): 24–47.

Euclid. *The Thirteen Books of Euclid's Elements.* Translated from the text of Heiberg, with introduction and commentary by Sir Thomas L. Heath. 2d. ed. rev. with additions. 3 vols. New York: Dover Publications, 1956.

Farrar, Frederic William, ed. *Essays on a Liberal Education.* London: Macmillan and Co., 1868.

Foote, George A. "The Place of Science in the British Reform Movement, 1830–1850." *Isis* 42 (1951): 192–208.

Foote, George A. "Science and its Function in Early Nineteenth Century England." *Osiris* 11 (1954): 438–54.

Forsyth, A. R. "Old Tripos Days at Cambridge." *The Mathematical Gazette* 19 (1935): 162–79.

Freudenthal, Hans. "The Main Trends in the Foundations of Geometry in the 19th Century." *Logic, Methodology and Philosophy of Science.* Edited by Ernest Nagel, Patrick Suppes and Alfred Tarski. Stanford: Stanford University Press, 1962. 613–21.

Freudenthal, Hans. "Zur Geschichte der Grundlagen der Geometrie." *Nieuw Archief voor Wiskunde* 4 (1957): 105–42.

Gauss, Carolo Friderico. *Disquisitiones generales circa superficies curvas.* Commentationes societatis regiae scientiarum Gottingensis recentiores 6, 1828. Reprinted in *Carl Friedrich Gauss Werke,* 12 vols. (Königlichen Gesellschaft der Wissenshaften zu Göttingen, 1880) vol. 4, pp. 217–58.

Gauss, Carlo Friderico and H. C. Schumacher. *Briefwechsel zwischen K. F. Gauss und H. C. Schumacher.* Altona: C. A. F. Peters, 1860–65.

Giere, Ronald N. and Richard S. Westfall, eds. *Foundations of Scientific Method: The Nineteenth Century.* Bloomington: Indiana University Press, 1973.

Grattan-Guiness, Ivar. "A Mathematical Union: William Henry and Grace Chisholm Young." *Annals of Science* 29 (1972): 105–86.

Grattan-Guiness, Ivar. "University Mathematics at the Turn of the Century: Unpublished Recollections by W. H. Young," *Annals of Science* 28 (1973): 369–84.

Graves, Robert Percival. *Life of Sir William Rowan Hamilton.* Dublin: Hodges, Figgis & Co., 1882–1889.

Gray, Jeremy. *Ideas of Space.* New York: Oxford University Press. (Clarendon), 1980.

Gray, Jeremy. "Non-Euclidean Geometry—A Re-Interpretation." *Historia Mathematica* 6 (1979): 236–58.

Gray, Jeremy and Laura Trilling. "Johann Heinrich Lambert; Mathematician and Scientist, 1728–1777." *Historia Mathematica* 5 (1978): 43–47.

Great Britain, Parliament. *Parliamentary Papers, 1852–53.* vol. 14 (*Reports from Commissioners,* vol. 5). Cmnd 2559, "Report of the Commissioners appointed to inquire into the State, Discipline, Studies and Revenues of the University and Colleges of Cambridge; together with the Evidence and an Appendix and Index."

Great Britain, Parliament. *Parliamentary Papers, 1864.* vol. 15. (*Reports from Commissioners,* vol. 6). "Report of Her Majesty's Commissioners Appointed to Inquire into the Revenues and Management of Certain Colleges and Schools, and the Studies pursued and Instruction given therein; with an Appendix and Evidence."

Great Britain, Parliament. *Parliamentary Papers, 1867–68.* vol. 15. (*Reports from Committees,* vol. 10), "Report from the Select Committee on Scientific Instruction; with the Proceedings of the Committee, Minutes of Evidence, and Appendix."

Gregory, Duncan Farquharson. *The Mathematical Writings of Duncan Farquharson*

Gregory. Edited by William Walton. Biographical Memoir by Robert Leslie Ellis. Cambridge: Deighton, Bell and Co., 1865.

Gregory, Duncan Farquharson. "On the Existence of Branches of Curves in Several Planes." *The Cambridge Mathematical Journal* 1 (1837–39): 259–66.

Griffin, Nicholas. "The Tiergarten Programme." *Russell* (forthcoming).

Haines IV, George. *Essays on German Influence upon English Education and Science, 1850–1919.* Connecticut College Monograph, no. 9. Connecticut College in association with Archon Books, 1969.

Haines IV, George. *German Influence upon English Education and Science, 1800–1866.* Connecticut College Monograph, no. 6. Connecticut College in association with Archon Books, 1957.

Halsted, George Bruce. "Bibliography of Hyperspace and Non-Euclidean Geometry." *American Journal of Mathematics* 1 (1878): 261–76, 384–5.

Halsted, George Bruce. "Bibliography of Hyperspace and Non-Euclidean Geometry." *American Journal of Mathematics* 2 (1879): 65–70.

Halsted, George Bruce. "Darwinism and Non-Euclidean Geometry." *Fiziko-matematicheskoe obshchestvo* 6 (1896): 22–25.

Halsted, George Bruce. "Light from Non-Euclidean Spaces on the Teaching of Elementary Geometry." *Scientiae Baccalaureus; A Quarterly Journal of Scientific Research* 1 (1890): 255–60.

Halsted, George Bruce. "Our Belief in Axioms, and the New Spaces." *Scientiae Baccalaureus; A Quarterly Journal of Scientific Research* 1 (1890); 11–19.

Hamilton, William. "Review of *Thoughts on the Study of Mathematics as a part of Liberal Education.*" *Edinburgh Review* 62 (1836): 218–52.

Hamilton, William Rowan. "On Symbolical Geometry." *The Cambridge and Dublin Mathematics Journal* 1 (1846): 45–57, 137–53, 256–63.

Hamilton, William and William Whewell. "Note to the Article on the Study of Mathematics in no. 126." *Edinburgh Review* 63 (1836): 142–44.

Hankins, Thomas L. "Algebra as Pure Time: William Rowan Hamilton and the Foundations of Algebra." *Motion and Time, Space and Matter: Interrelations in the History of Philosophy and Science.* Edited by Peter K. Machamer and Robert G. Turnbull. Columbus: Ohio State University Press, 1978. 327–59.

Hankins, Thomas L. *Sir William Rowan Hamilton.* Baltimore: Johns Hopkins University Press, 1978.

Hankins, Thomas L. "Triplets and Triads: Sir William Rowan Hamilton on the Metaphysics of Mathematics." *Isis* 68 (1977): 175–93.

Hardy, G[odfrey] H[arold]. *A Mathematician's Apology.* With a foreword by C. P. Snow. Cambridge: University Press, 1982.

Helmholtz, Hermann von. "The Axioms of Geometry." *The Academy* 3 (1872): 52–53.

Helmholtz, Hermann von. "The Axioms of Geometry." *The Academy* 1 (1870): 128–31.

Helmholtz, Hermann von. "The Origin and Meaning of Geometrical Axioms." *Mind* 1 (1876): 301–21.

Helmholtz, Hermann von. "The Origin and Meaning of Geometrical Axioms (II)." *Mind* 3 (1878): 212–25.

Helmholtz, Hermann von. "Sur les faits qui servent de base a la géométrie." Translated by J. Houël. *Societé des sciences physiques et naturelles de Bordeaux,* series 1, 5 (1867): 372–78.

Helmholtz, Hermann von. "Über die Thatsachen, die der Geometrie zum Grunde liegen." *Nachrichten von der königlichen Gesselschaft der Wissenschaften zu Göttingen* (1868): 193–222.

Helmholtz, Hermann von. "Über die thatsälichen Grundlagen de Geometrie." *Verhandlungen des naturhistorisch-medicinischen Vereins zu Heidelberg* 4 (1866): 197–202.

Henrici, Olaus. "Presidential Address: Mathematics and Physics Section." *Report of the Fifty-third Meeting of the BAAS held at Southport in September 1883.* London: John Murray, 1884. 393–400. Reprinted in *Nature* (1883): 497–500.

Henrici, Olaus. "s.v. Geometry." *Encyclopaedia Britannica,* 9th ed. (1883).

Herschel, John Frederick William. *Essays from the Edinburgh and Quarterly Reviews with Addresses and Other Pieces.* London: Longman, Brown, Green, Longmans, & Roberts, 1857.

Herschel, John Frederick William. *Herschel at the Cape: Diaries and Correspondence of Sir John Herschel, 1834–1838.* Edited with an introduction by David S. Evans, Terrence J. Deening, Betty Hall Evans and Stephen Goldfarb. Austin: University of Texas Press, 1969.

Herschel, John Frederick William. *A Preliminary Discourse on the Study of Natural Philosophy.* London, 1830. Reprint. New York: Johnson Reprint Corporation, 1966.

Herschel, John Frederick William. "Presidential Address." *Report of the Fifteenth Meeting of the BAAS Held at Cambridge in June 1845.* London: John Murray, 1846. xxvii–xliv.

Herschel, John Frederick William. "Review of Whewell's *History of the Inductive Sciences* (1837) and *Philosophy of the Inductive Sciences* (1840)." *The Quarterly Review* 68 (1841): 177–238.

Herschel, John Frederick William. *Sir John Herschel and Education at the Cape, 1834–1840.* Compiled by W. T. Ferguson in collaboration with R. F. M. Immelman. Cape Town: Oxford University Press, 1961.

"Higher Education in Germany." (article signed G. S.) *Nature* 75 (1907): 338.

Hill, Thomas. *Geometry and Faith: A Fragmentary Supplement to the Ninth Bridgewater Treatise.* 2nd ed. revised and enlarged. New York: G. P. Putnam's Sons, 1874.

Houghton, Walter E. *The Victorian Frame of Mind, 1830–1870.* New Haven: Yale University Press, 1957.

Huxley, Leonard. *Life and Letters of Thomas Henry Huxley.* London: Macmillan and Co., 1913.

Huxley, Thomas H. "The Scientific Aspects of Positivism." *The Fortnightly Review* ns. 30 (1869): 653–70.

Huxley, Thomas H. "Scientific Education: Notes of an After Dinner Speech." *Macmillan's Magazine* 20 (1869): 177–84.

Hyman, Anthony. *Charles Babbage: Pioneer of the Computer.* Princeton: Princeton University Press, 1982.

Ingleby, C. M.   "The Antinomies of Kant." *Nature* 7 (1873): 262.
Ingleby, C. M.   "Prof. Clifford on Curved Space." *Nature* 7 (1873): 282.
Ingleby, C. M.   "Transcendent Space." *Nature* 1 (1869–70); 289, 361.
[James Fitzjames Stephen].   *Essays by a Barrister.* London: Smith, Elder and Co., 1862.
Jammer, Max.   *Concepts of Space: The History of Theories of Space in Physics.* Foreword by Albert Einstein. Cambridge: Harvard University Press, 1954.
Jensen, J. Victor.   "The X-Club: Fraternity of Victorian Scientists." *The British Journal for the History of Science* 4 (1970): 63–72.
Jevons, William Stanley.   "Helmholtz on the Axioms of Geometry." *Nature* 4 (1871): 481–82.
Jevons, William Stanley.   "John Stuart Mill's Philosophy Tested." *The Contemporary Review* 31 (1877–78); 167–82, 256–75.
Jevons, William Stanley.   *Letters and Journal of W. Stanley Jevons.* Edited by his wife. London: Macmillan and Co, 1886.
Jevons, William Stanley.   *The Principles of Science: A Treatise on Logic and Scientific Method.* Introduction by Ernest Nagel. New York: Dover Publications, 1958.
Jevons, William Stanley.   *Pure Logic and Other Minor Works.* Edited by Robert Adamson and Harriet A. Jevons. Preface by Robert Adamson. London: Macmillan and Co., 1890.
Jones, E. E. Constance.   "Letter." *Nature* 47 (1892): 78.
Joseph Agassi.   "Sir John Hershel's Philosophy of Success." *Historical Studies in the Physical Sciences* 1 (1969): 1–36.
[Kant, Immanuel].   *Kant's Inaugural Dissertation and Early Writings on Space.* Translated by John Handyside. Chicago: Open Court Publishing Co., 1929.
Kelland, Professor [John].   "On the Limits of our Knowledge Respecting the Theory of Parallels." (Read Dec. 21, 1863). *Transactions of the Royal Society of Edinburgh* 23 (1861–64): 433–50.
Kitcher, Philip.   *The Nature of Mathematical Knowledge.* New York: Oxford University Press, 1983.
Klein, Felix.   *Elementary Mathematics from an Advanced Standpoint.* Translated from the Third German Edition by E. R. Hedrich and C. A. Noble. New York: Macmillan Company, 1939.
Klein, Felix.   *The Evanston Colloquium: Lectures on Mathematics Delivered from August 28 to September 9, 1893, before Members of the Congress of Mathematics held in Connection with the World's Fair in Chicago at Northwestern University, Evanston, Ill.* Reported by Alexander Ziwet. New York: Macmillan and Co., 1894.
Klein, Felix.   *Gesammelte Mathematische Abhandlungen.* 3 vols. Berlin: Springer, 1921–23.
Klein, Felix.   *Nicht-Euklidische Geometrie: Vorlesung Gehalten Während des Wintersemesters 1889–90 and Sommersemesters 1890.* Edited by Fr. Schilling. 2 vols. Göttingen, 1893.
Klein, Felix.   *Le programme d'Erlangen.* Preface de J. Dieudonné. Postface de P. François Russo. Paris; Gauthier-Villars, 1974.
Klein, Felix.   "Riemann et son influence ser les mathématiques modernes."

Oeuvres mathematiques de Riemann. Translated by L. Laugel. Preface by M. Hermite. Paris: Albert Blanchard, 1968.

Klein, Felix. "Ueber die sogenannte Nicht-Euklidische Geometrie." *Mathematische Annalen* 4 (1871): 573–625.

Klein, Felix. "Ueber die sogenannte Nicht-Euklidische Geometrie." *Mathematische Annalen* 6 (1873): 112–45.

Klein, Felix. "Ueber die sogennante Nicht-Euklidische Geometrie." *Nachrichten von der Königlichen Gesellschaft der Wissenschaften zu Göttingen* 17 (1871): 419–33.

Kline, Morris. *Mathematical Thought from Ancient to Modern Times.* New York: Oxford University Press, 1972.

Knott, Cargill Gilston. *Life and Scientific Work of Peter Guthrie Tait.* Cambridge: University Press, 1911.

Koenigsberger, Leo. "The Investigations of Hermann von Helmholtz on the Fundamental Principles of Mathematics and Mechanics." *Smithsonian Institution Annual Reports* (1898): 93–124.

Koppelmann, Elaine. "The Calculus of Operations and the Rise of Abstract Algebra." *Archive for History of Exact Sciences* 8 (1971): 155–242.

Kuhn, Thomas. *The Structure of Scientific Revolutions.* Chicago: University of Chicago Press, 1962.

[Langley E. M.]. E. M. L. "Review of Dixon's *The Foundations of Geometry.*" *Nature* 43 (1891): 554–55.

Land, J. P. N. "Kant's Space and Modern Mathematics." *Mind* 2 (1877); 38–46.

Layton, David. "The Educational Work of the Parliamentary Committee of the British Association for the Advancement of Science." *History of Education* 5 (1976): 25–39.

Layton, David. *Science for the People: The Origins of the School Science Curriculum in England.* New York: Science History Publications, 1973.

The Leibniz-Clarke Correspondence. Edited with introduction and notes by H. G. Alexander. Manchester: Manchester University Press, 1956.

Lewes, George Henry. Appendix to "Imaginary Geometry and the Truth of Axioms." *Problems of Life and Mind.* First Series, "The Foundations of a Creed." vol. 2. London: Tubner & Co., 1875.

Lewes, George Henry. "Kant's View of Space." *Nature* 1 (1870): 289, 334.

Lobachevskii, N. I. *Etudes géométriques sur la théorie des parallèles, suivi d'un extrait de la correspondence de Gauss et de Schumacher.* Translated by J. Houël, Paris: Gauthier-Villars, 1866.

Lockyer, Norman. "Presidential Address." *Report of the Seventy-third Meeting of the BAAS Held at Southport in September 1903.* London: John Murray, 1904.

Loria, Gino. "A Few Remarks on the 'Syllabus of Modern Plane Geometry' Issued by the AIGT." *Eighteenth Annual Report of the Association for the Improvement of Geometrical Teaching.* Bedford: W. J. Robinson, 1893. 49–53.

Loria, Gino. "Sketch of the Origin and Development of Geometry prior to 1850." Translated by G. B. Halsted. *Monist* 13 (1902–03): 80–102, 218–34.

Losee, J. "Whewell and Mill on the Relations between Philosophy of Science and

History of Science." *Studies in the History and Philosophy of Science* 14 (1979): 113–26.

Lowe, Victor. *Alfred North Whitehead: The Man and His Work.* vol. 1. Baltimore: Johns Hopkins University Press, 1985.

Macfarlane, Alexander. *Lectures on Ten British Mathematicians of the Nineteenth Century.* New York: Wiley, 1916.

MacLeod, Roy M. "A Note on *Nature* and the Social Significance of Scientific Publishing, 1850–1914." *Victorian Periodicals Newsletter* 3 (1968): 16–17.

MacLeod, Roy M. "Resources of Science in Victorian England: The Endowment of Science Movement, 1868–1900." *Science and Society 1600–1900.* Edited by Peter Mathias. Cambridge: University Press, 1972. 111–66.

MacLeod, Roy M. "The X-Club: A Social Network of Science in Late-Victorian England." *Notes and Records of the Royal Society of London* 24 (1969): 305–22.

MacLeod, Roy M., and Russell Moseley. "The Naturals and Victorian Cambridge: Reflections on the Anatomy of an Elite." *Oxford Review of Education* 6 (1980): 177–95.

MacPherson, Robert Greer. *Theory of Higher Education in Nineteenth Century England.* Athens: University of Georgia Press, 1959.

Maxwell, James Clerk. *Matter and Motion.* Reprinted with notes and appendices by Sir Joseph Larmor. New York: Dover Publications, 1952.

Merz, John Theodore. *A History of European Thought in the Nineteenth Century.* 4 vols. London: W. Blackwood and Sons, 1912–28.

Mill, John Stuart. *Autobiography of John Stuart Mill.* With a preface by John Jacob Coss. New York: Columbia University Press, 1924.

Mill, John Stuart. *An Examination of Sir William Hamilton's Philosophy and of the Principle Philosophical Questions discussed in his Writings.* 2nd ed. London: Longmans, Green, and Co., 1865.

Mill, John Stuart. *The Philosophy of Scientific Method.* Edited with an Introduction by Ernest Nagel. New York: Hafner, 1950.

Mill, John Stuart. *A System of Logic Ratiocinative and Inductive: Being a Connected View of the Principles of Evidence and the Methods of Scientific Investigation.* New York: Harper & Brothers, 1846.

Monge, Gaspard. *Géométrie descriptive: Leçons données aux écoles normales, l'an 3 de la République.* Paris: Boudouin, an viii [1798].

Morrell, J. B. "Individualism and the Structure of British Science in 1830." *Historical Studies in the Physical Sciences* 3 (1971): 183–204.

Morrell, Jack, and Arnold Thackray. *Gentlemen of Science: Early Years of the British Association for the Advancement of Science.* Oxford: Clarendon Press, 1981.

Moseley, Maboth. *Irascible Genius: A Life of Charles Babbage, Inventor.* Foreword by B. V. Bowden. London: Hutchinson, 1964.

Muirhead, R. F. "The Teaching of Mathematics." *Mathematical Gazette* 2 (1901): 81–83.

Mulcahy, John. *Principles of Modern Geometry with Numerous Applications to Plane and Spherical Figures.* 2nd ed. revised. Dublin: Hodges and Smith, 1862.

Nagel, Ernest. "The Formation of Modern Conceptions of Formal Logic in the Development of Geometry." *Osiris* 7 (1939): 142–22.

Nagel, Ernest. "'Impossible Numbers': A Chapter in the History of Modern Logic." *Studies in the History of Ideas.* Edited by the Department of Philosophy of Columbia University. vol. 3. New York: Columbia University Press, 1935. 429–74.

Novy, Lubos. *Origins of Modern Algebra.* Translated by Jaroslav Tauer. Leyden: Noordhoff Publishing Co., 1973.

Olson, Richard. *Scottish Philosophy and British Physics, 1750–1880: A Study in the Foundations of the Victorian Scientific Style.* Princeton: Princeton University Press, 1975.

Olson, Richard. "Scottish Philosophy and Mathematics: 1750–1830." *Journal of the History of Ideas* 32 (1971): 29–44.

Pasch, Moritz. *Vorlesungen über neure geometrie.* Leipzig: B. G. Teubner, 1882.

Passmore, John. *A Hundred Years of Philosophy.* rev. ed. New York: Basic Books, 1966.

Peacock, George. *Observations on the Statutes of the University of Cambridge.* London: J. W. Parker, 1841.

Pearson, Karl. *The Grammar of Science.* London: W. Scott, 1892.

Pearson, Karl. "Old Tripos Days at Cambridge, as Seen from Another Viewpoint." *The Mathematical Gazette* 20 (1936): 27–36.

Perry, John. "The Mathematical Tripos at Cambridge." *Nature* 75 (1907): 273–74.

Perry, John. "Mathematics in the Cambridge Locals." *Nature* 67 (1902): 81–82.

Perry, John. *Discussion on the Teaching of Mathematics. To Which is now added the Report of the British Association Committee drawn up by the Chairman, Professor Forsyth.* British Association Meeting at Glasgow, 1901. London: Macmillan and Co., 1902.

Plücker, Julius. "On a New Geometry of Space." *Philosophical Transactions of the Royal Society* 155 (1865); 725–91.

Poincaré, Henri. *Dernières Pensées.* Paris: E. Flammarion, 1913.

Poincaré, Henri. *The Foundations of Science; Science and Hypothesis, The Value of Science, Science and Method.* Authorized translation by George Bruce Halsted; with a special preface by Poincaré and an introduction by Josiah Royce. New York: Science Press, 1913.

Poincaré, Henri. "Non-Euclidean Geometry." Translated by W. J. L. *Nature* 45 (1892).

Pycior, Helena M. "Early Criticisms of the Symbolical Approach to Algebra." *Historia Mathematica* 9 (1982): 413–40.

Pycior, Helena M. "George Peacock and the British Origins of Symbolical Algebra." *Historia Mathematica* 8 (1981): 23–45.

Pycior, Helena M. "The Three Stages of Augustus De Morgan's Algebraic Work." *Isis* 74 (1983): 211–22.

Reynolds, E. M. *Modern Methods in Elementary Geometry.* London: Macmillan and Co., 1868.

Rhodes, E. Hawksley. "The Postulates of the Science of Space." *The Academy* (1875): 352.

Richards, Joan L. "The Art and the Science of British Algebra: A Study in the Perception of Mathematical Truth." *Historia Mathematica* 7 (1980): 343–65.

Richards, Joan L.   "Augustus De Morgan, the History of Mathematics, and the Foundations of Algebra." *Isis* 78 (1987): 7–30.

Richards, Joan L.   "The Evolution of Empiricism: Hermann von Helmholtz and the Foundations of Geometry." *British Journal for the Philosophy of Science* 28 (1977): 235–53.

Richards, Joan L.   "Projective Geometry and Mathematical Progress in Mid-Victorian Britain." *Studies in the History and Philosophy of Science* 17 (1968): 297–325.

Richards, Joan L.   "The Reception of a Mathematical Theory: Non-Euclidean Geometry in England, 1868–1883." *Natural Order: Historical Studies of Scientific Culture.* Barry Barnes and Stephen Shapin, eds. Beverly Hills: Sage Publications, 1979. 143–66.

Riemann, Bernhard.   "On the Hypotheses which Lie at the Bases of Geometry." Translated by W. K. Clifford. *Nature* 8 (1873): 14–17, 36–37.

Riemann, Bernhard.   "Über die Hypotheses, welche der Geometrie zu Grunde liegen." *Abhandlungen der Königlichen Gesellschaft der Wissenschaften zu Göttingen* 13 (1867): 132–52.

Roberts, Samuel.   "Remarks on Mathematical Terminology, and the Philosophic Bearing of Recent Mathematical Speculations concerning the Realities of Space." (Read Nov. 9, 1882). *Proceedings of the London Mathematical Society* 14 (1882–83): 5–15.

Rothblatt, Sheldon.   *The Revolution of the Dons: Cambridge and Society in Victorian England.* New York: Basic Books, 1968.

Rothblatt, Sheldon.   *Tradition and Change in English Liberal Education.* London: Faber and Faber, 1976.

Russell, Bertrand.   *The Autobiography of Bertrand Russell, 1872–1914.* Boston: Little, Brown and Co., 1967.

Russell, Bertrand.   *Collected Papers of Bertrand Russell.* Edited by Kenneth Blackwell et al. vol. 1 and following. London: George Allen and Unwin, 1983–.

Russell, Bertrand.   *An Essay on the Foundations of Geometry.* Cambridge: University Press, 1897.

Russell, Bertrand.   *My Philosophical Development.* London: George Allen and Unwin, 1959.

Russell, Bertrand.   "Observations on Space and Geometry." The Bertrand Russell Archives, McMaster University, Hamilton, Ontario. 1895.

Russell, Bertrand.   "Sur les axiomes de la géométrie." *Revue de métaphysique et de morale* 7 (1899): 684–707.

Russell, Bertrand.   "A Turning-Point in My Life." *The Saturday Book* 8 (1948): 142–46.

[Saccheri, Girolamo].   *Girolamo Saccheri's Euclides Vindicatus.* Edited and translated by George Bruce Halsted. Chicago: Open Court, 1920.

Salmon, George.   "On Some Points in the Theory of Elimination." *The Quarterly Journal of Pure and Applied Mathematics* 7 (1866); 327–37.

Salmon, George.   *A Treatise on Conic Sections.* 5th ed. London: Longmans, Green, Reader, and Dyer, 1869.

Salmon, George. *A Treatise on the Higher Plane Curves*. Dublin: Hodges and Smith, 1852.

Schweber, Silvan S. "Auguste Comte and the Nebulous Hypothesis." forthcoming.

Schweber, Silvan S. "Scientists as Intellectuals: The Early Victorians." *Victorian Science and Victorian Values*. James Paradis and Thomas Postlewait, eds. New York Academy of Sciences Annals, vol. 360, 1981. 1–37.

Smith, David Eugene. "Euclid, Omar Khayyam, and Saccheri." *Scripta Mathematica* 3 (1935): 5–10.

Smith, H. J. Stephen. "On the Present State and Prospects of Some Branches of Pure Mathematics." (Read Nov. 9, 1876). *Proceedings of the London Mathematical Society* 8 (1876–77): 6–29.

Sommerville, Duncan M. Y. *Bibliography of Non-Euclidean Geometry, including the Theory of Parallels, the Foundation of Geometry, and Space of n-Dimensions*. London: Harrison & Sons, 1911.

Spottiswoode, William. "Address to Mathematics and Physics Section." *Report of the Thirty-fifth Meeting of the BAAS held at Birmingham in September, 1865*. London: John Murray, 1866. 1–6.

Spottiswoode, William. "Presidential Address." *Report of the Forty-eighth Meeting of the BAAS Held at Dublin in August, 1878*. London: John Murray, 1879. 1–32.

Staudt, Karl Georg Christian von. *Geometrie der Lage*. Nürnberg: Bauer und Raspe, 1847.

Stephen, J[ames] F[itzjames]. "Necessary Truth." *The Contemporary Review* 25 (1874–75): 44–73.

Strong, E. W. "Whewell vs. J. S. Mill on Science." *Journal of the History of Ideas* 16 (1955); 209–31.

Sylvester, James J. *The Collected Mathematical Papers of James Joseph Sylvester*. Cambridge: University Press, 1904–12.

Sylvester, James J. "Presidential Address: Mathematics and Physics Section." *Report of the Thirty-ninth Meeting of the BAAS held at Exeter in August, 1869*. London: John Murray, 1870. 1–9. Revised and reprinted as "A Plea for the Mathematician." *Nature* 1 (1869–70): 237–39, 260–63.

Taylor, Charles. *An Introduction to the Ancient and Modern Geometry of Conics*. Cambridge: Deighton, Bell, and Co., 1881.

[Thompson, Thomas Perronet]. *Geometry without Axioms, or the First Book of Euclid's Elements*. By a member of the University of Cambridge. 5th ed. London: R. Heward, 1934.

Todhunter, Isaac. *The Conflict of Studies and other Essays on Subjects Connected with Education*. London: Macmillan and Co., 1873.

Todhunter, Isaac. *William Whewell, D. D., Master of Trinity College, Cambridge. An Account of his Writings with Selections from his Literary and Scientific Correspondence*. London: Macmillan and Co., 1876.

Torretti, Roberto. *Philosophy of Geometry from Riemann to Poincaré*. Boston: D. Reidel, 1978.

Toth, Imre. "Gott und Geometrie: Eine viktorianische Kontroverse." *Evolutionstheorie und ihre Evolution: Vortragsreihe der Universität Regensburg zum 100 Todestag von Charles Darwin.* Dieter Heinrich, ed. Schriftenreihe der Universität Regensburg. band 7, 1982. 141–204.

Toth, Imre. "La Revolution non-Euclidean." *La Recherche en histoire des sciences* 75 (1983): 241–65.

Toth, Imre. "Langage et pensée mathematiques." Paper delivered at Seminaire de Mathématik du Centre Universitaire de Luxembourg. June 9–11, 1976.

Tupper, J. L. "Prof. Helmholtz and Prof. Jevons." *Nature* 5 (1872): 202–203.

Turner, Frank Miller. *Between Science and Religion: The Reaction to Scientific Naturalism in Late Victorian England.* New Haven: Yale University Press, 1974.

Turner, Frank Miller. "The Victorian Conflict between Science and Religion: A Professional Dimension." *Isis* 69 (1978): 356–76.

Walton, William. "On the Doctrine of Impossibles in Algebraic Geometry." *The Cambridge and Dublin Mathematical Journal* 7 (1852): 234–42.

W[alton], W[illiam]. "On the Existence of Possible [Real] Asymptotes to Impossible [Imaginary] Branches of Curves." *The Cambridge Mathematical Journal* 2 (1840): 236–39.

Walton, William. "On the General Interpretation of Equations between Two Variables in Algebraic Geometry." *The Cambridge Mathematical Journal* 2 (1840): 103–13.

Walton, William. "On the General Theory of Multiple Points." *The Cambridge Mathematical Journal* 2 (1840): 155–67.

Walton, William. "On the General Theory of the Loci of Curvilinear Intersection." *The Cambridge Mathematical Journal* 2 (1840): 85–91.

Ward, William. "Necessary Truth (In Answer to Mr. Fitzjames Stephen)." *The Contemporary Review* 25 (1874–75): 527–45.

Ward, William. "A Reply on Necessary Truth." *The Dublin Review* 23 (1874): 54–63.

Watson, E. C. "College Life at Cambridge in the Days of Stokes, Cayley, Adams and Kelvin." *Scripta Mathematica* 6 (1939): 101–106.

Whewell, William. *Astronomy and General Physics Considered with Reference to Natural Theology.* London: W. Pickering, 1833.

Whewell, William. *The Doctrine of Limits, with its Applications.* Cambridge: J. and J. J. Deighton, 1838.

Whewell, William. *History of the Inductive Sciences from the Earliest to the Present Time.* 3rd ed. 3 vols. London: John W. Parker and Son, 1837. Reprint. London: Frank Cass & Co., 1967.

Whewell, William. *Indications of the Creator. Extracts Bearing upon Theology, from the History and the Philosophy of the Inductive Sciences.* London: J. W. Parker, 1845.

Whewell, William. *Influence of the History of Science upon Intellectual Education. A Lecture before the Royal Institution of Great Britain.* Boston: Gould and Lincoln, 1854.

Whewell, William. *The Mechanical Euclid, to which are added Remarks on Mathemat-*

ical Reasoning and on the Logic of Induction. Cambridge: J. and J. J. Deighton, 1837.

Whewell, William. *Of a Liberal Education in General; and with Particular Reference to the Leading Studies of the University of Cambridge.* London: J. W. Parker, 1845.

Whewell, William. "On the Fundamental Antithesis of Philosophy." (Read Feb. 5, 1844). *Transactions of the Cambridge Philosophical Society* 8 (1849): 170–81.

Whewell, William. *On the Principles of English University Education.* 2nd ed. London: John W. Parker, 1838.

Whewell, William. *The Philosophy of the Inductive Sciences.* new ed. 2 vols. London: John W. Parker, 1847. Reprint. London: Frank Cass & Co., 1967.

Whitehead, Alfred North. *Essays in Science and Philosophy.* New York: Philosophical Library, 1947.

Whitehead, Alfred North. *The Organisation of Thought, Educational and Scientific.* London: William and Norgate, 1917.

Whitehead, Alfred North. "Presidential Address to the Mathematics and Physics Section: The Organisation of Thought." *Report of the Eighty-sixth Meeting of the BAAS Held at Newcastle-on-Tyne in September, 1916.* London: John Murray, 1917.

Whitehead, Alfred North. "s.v. Axioms of Geometry." *Encyclopaedia Britannica,* 11th ed. (1910–11).

Whittaker, E. T. "A Catalogue Raisonné of Sir Robert Ball's Mathematical Papers." *Reminiscences and Letters of Sir Robert Ball.* Edited by W. Valentine Ball. Boston: Little, Brown and Co., 1915.

Wiegand, Sylvia. "Grace Chisholm Young." *Association for Women in Mathematics Newsletter* 7 (1977): 5–10.

Willey, Basil. *More Nineteenth Century Studies: A Group of Honest Doubters.* London: Chatto & Windus, 1956.

Willey, Basil. *Nineteenth Century Studies: Coleridge to Matthew Arnold.* London: Chatto & Windus, 1949.

Williams, Leslie Pearce. *Michael Faraday, A Biography.* New York: Basic Books, 1965.

Wilson, James Maurice. *Elementary Geometry.* London: Macmillan and Co., 1868.

Wilson, James Maurice. "Euclid as a Text-Book of Elementary Geometry." *The Educational Times and Journal of the College of Preceptors* 21 (1868): 126–28.

Wilson, James Maurice. "Reply to [De Morgan's] Review." *The Athenaeum* 2129 (1868): 216.

Winstanley, Denys Arthur. *Early Victorian Cambridge.* Cambridge: University Press, 1940.

Winstanley, Denys Arthur. *Later Victorian Cambridge.* Cambridge: University Press, 1947.

Wolff, Ferdinand. "Critical Examination of Euclid's First Principles compared to Those of Modern Geometry." *The Quarterly Journal of Pure and Applied Mathematics* 8 (1867): 301–309.

Wolff, Ferdinand. "Critical Examination of the First Propositions of Euclid." *The Quarterly Journal of Pure and Applied Mathematics* 7 (1866): 76–81.

Youmans, E[dward] L[ivingston]. *The Culture Demanded by Modern Life; A Series of Addresses and Arguments on the Claims of Scientific Education.* New York: D. Appleton & Co., 1867.

Young, George Malcom. *Victorian England: Portrait of an Age.* London: Oxford University Press, 1936.

Young, Robert M. "The Impact of Darwin on Conventional Thought." *The Victorian Crisis of Faith.* Edited by Anthony Symondson. London: Camelot Press, 1970. 13–35.

Ziegler, Renatus. *Die Geschichte der Geometrischen Mechanik im 19. Jahrhundert.* Stuttgart: Franz Steiner Verlag Weisbaden, 1985.

INDEX